10 Springer Series in Solid-State Sciences
Edited by Manuel Cardona and Peter Fulde

Springer Series in Solid-State Sciences

Editors: M. Cardona P. Fulde H.-J. Queisser

H. Bilz   W. Kress

Phonon Dispersion Relations in Insulators

With 162 Figures in 271 Separate Illustrations

Springer-Verlag Berlin Heidelberg New York 1979

Professor Dr. Heinz Bilz
Dr. Winfried Kress

Max-Planck-Institut für Festkörperforschung
Heisenbergstrasse 1, D-7000 Stuttgart 80, Fed. Rep. of Germany

Series Editors:

Professor Dr. Manuel Cardona
Professor Dr. Peter Fulde
Professor Dr. Hans-Joachim Queisser

Max-Planck-Institut für Festkörperforschung
Heisenbergstrasse 1, D-7000 Stuttgart 80, Fed. Rep. of Germany

ISBN-13: 978-3-642-81349-8   e-ISBN-13: 978-3-642-81347-4
DOI: 10.1007/978-3-642-81347-4

Library of Congress Cataloging in Publication Data. Bilz, Heinz, 1926-. Phonon dispersion relations in insulators. (Springer series in solid-state sciences ; 10). Bibliography: p. Includes index. 1. Crystals, Effect of radiation on--Charts, diagrams, etc. 2. Phonons–Scattering–Charts, diagrams, etc. I. Kress, Winfried, 1941- joint author. II. Title. III. Series. QC176.8.R3B54   530.4'1   79-17304

© by Springer-Verlag Berlin Heidelberg 1979
Softcover reprint of the hardcover 1st edition 1979

2153/3130-5 4 3 2 1 0

Preface

This phonon atlas presents a collection of phonon-dispersion and density-of-states curves of more than a hundred insulating crystals. It grew out of an appendix to a handbook article on phonon spectra [2.1] from which it was finally separated mainly because this phonon atlas provides a rather self-contained tool for every scientist who is working in the field of dynamical properties of solids. He often may find it' useful to have a handy documentation of the experimental phonon dispersion curves which have been measured so far, together with information on calculated dispersion relations and densities of states.

The book will be found to be incomplete by readers who are interested not only in phonon frequencies of a specific crystal but would also like to know about related properties such as elastic and dielectric constants. This is, at the present time, beyond the scope of this volume, but the authors would welcome all suggestions and criticism which could be considered for a forthcoming edition. Furthermore, we would be pleased to provide interested readers with information about phonon spectra which came to our knowledge after completion of the manuscript. On the other hand, we will be most grateful for all information about phonon dispersion curves which is missing in our collection or new data for further editions.

During the early stage of this work, Dr. Erika Kiefer-Schröder and Professor Ulrich Schröder gave us important help in the setup of this documentation and its realization. The authors are most grateful for their contributions. In addition, the authors have been encouraged by many colleagues to undertake this attempt of phonon documentation. We are indebted to all of them. The FIZ 4 at Karlsruhe generously provided us with information retrieval runs. We thank them, in particular Dr. Behrends, Dr. Ebel, and Dipl. Phys. E. Zybell, for their support. One of the authors gratefully acknowledges the financial support by BMFT, Bonn. Finally, it is a real pleasure to thank Miss Gisela Keck for her careful and patient preparation of the manuscript.

Stuttgart, September 1979   H. Bilz   W. Kress

Contents

Part I

Summary of
Theory of Phonons

1. Introduction

One of the most powerful methods to determine phonon dispersion relations in
a crystal is inelastic neutron scattering which yields phonon frequencies of
all wavevectors with an accuracy of a few percent or better if circumstances
are favorable. Infrared and Raman spectra provide optic frequencies only at
very long wavelengths with an accuracy of about 0.1%. From these data, and
including information obtained from elastic and dielectric constants, it is
possible in many cases to calculate phonon dispersion relations which inter-
polate between the experimental data. In the simplest case, a set of inter-
ionic force constants is fitted, usually by a least-square procedure, to the
data in such a way as to give agreement within the limits of experimental
error. The number of force constants which is required in such a procedure
becomes, however, often rather high (≥ 20) even in simple cubic crystals.
Therefore, phonon models are widely used which start from a parametrized
"model" description of the electron-ion interaction before going, via the
adiabatic condition, to the "formal" force constants. This leads usually to
a drastic reduction of parameters (<10 for diatomic crystals) and facilitates
considerably the physical interpretation of the phonon spectra. A summary
of the most important models and a short discussion of the underlying physical
ideas is given in Part I.

In Part II dispersion curves and densities of states of more than 100
insulating crystals are represented. They are grouped into families mainly
determined by the crystal structure. In some cases a separate chapter is
devoted to a specific group of crystals, e.g., the alkali halides, which
has been particularly well investigated. The transition metal compounds
with rock salt structure which are in fact metals, are included (Chap. 6)
since they exhibit many interesting features related to insulating crystals
with rock salt structure although they show characteristic differences, too.
At the beginning of each chapter, the individual crystals of the family are
listed, together with a reference to the number of the figure in the text.
In addition an alphabetic list of crystals is given in the subject index at
the end of the book. The table on p.15 informs the reader about abbreviations

used in the figure captions. These captions contain, in a very condensed form, the necessary information about literature, thermodynamic data (temperature and pressure), the type of phonon model, and the number of parameters used in the calculation. The short notations under M: FCM, RIM etc. refer to corresponding subsections in Part I. Additional information can be found in the theoretical papers cited in the figure captions. For the majority of crystals, usually only room-temperature data are available. In a few cases, low-temperature measurements have been carried out and are reported in the text. In cases where new data have replaced less accurate older measurements, only the new results are represented; reference to the former data is usually made in the literature cited.

While selecting the dispersion curves of a crystal for the atlas, emphasis was put on the original experimental papers even if, occasionally, slightly better calculations were available. These are cited in the figure captions. In case of the densities-of-states curves, the most reliable model calculation was chosen except for the very few cases where direct measurements existed. The number of references cited for a given crystal was limited in order to prevent captions from becoming too cumbersome.

The decision about the theoretical calculations to be reported on in the text occasionally gave the authors a splitting headache because of conflicting arguments to be considered. On one side, the scientist usually looks for a description of data which is as precise as possible so that related entities, in particular densities of states, can be determined with high accuracy. On the other side, the theoretical models used in these types of careful fitting-to-the-data calculations are not always the best from a physical point of view. Sometimes, they contain a certain number of parameters which prevent useful extensions of the model to other entities such as Raman scattering, etc. We have decided in several cases where a simpler and "better" model (in the author's opinion) was nearly as good as a more accurate one to show the first version. There is always sufficient literature given in the captions to find out about further theoretical calculations. In one case (ZnO) the situation was found to be extremely controversial; therefore, two different versions are presented in the text. The authors hope that some "white spots" in this atlas may stimulate further experimental investigations and, quite generally, that low-temperature and high pressure data will become available to a larger extent during the coming decade.

It is felt that many of the users of this volume (aside from the few experts in the field) would welcome a short introduction to the different models; this is given in the following chapter. For more details, the reader is referred to the many textbooks and articles in the field. A selection is given in the references. We note that notation always follows closely the above-cited handbook article on phonon spectra [2.1].

2. Phonon Dispersion Relations and Phonon Models

2.1 Formal Force Constants (FCM)

The lattice potential ϕ of a crystal in thermal equilibrium can be derived from the free energy $F(T,V)$. ϕ reads in the classical harmonic approximation $(\phi \simeq \phi_2)$,

$$\phi_2 = \frac{1}{2} \sum_{L,L'} \underline{u}(L)\underline{\phi}(LL')\underline{u}(L') \tag{2.1}$$

where $L = (\ell,\kappa)$ denotes the κth particle in the ℓth cell. The $\underline{u}(\ell,\kappa)$ denote the displacement vectors of the ions with equilibrium positions, $\underline{x}(\ell,\kappa) = \underline{x}(\ell) + \underline{x}(\kappa)$, where $\underline{x}(\ell)$ is pointing to the ℓth cell of the crystal and $\underline{x}(\kappa)$ describes the relative position of the κth ion in the cell. The 3 × 3 two-ion force constant matrices $\underline{\phi}(L,L')$ are subject to the conservation laws of energy, momentum, and angular momentum and to the symmetry restrictions of the space group [2.2].

We note that our treatment is based on the assumption of crystal stability and on the adiabatic approximation (see [2.3]). Furthermore, the force constants being derived from a series expansion of the free energy $F(T,V)$ are pseudo-harmonic entities which contain a quasi-harmonic correction due to thermal expansion [2.4] as well as a self-energy renormalization arising from anharmonicity (see [2.1,5]).

From the vector equations of motion

$$M_\kappa \underline{\ddot{u}}(L) = - \frac{\partial \phi_2}{\partial \underline{u}(L)} = - \sum_{L'} \underline{\phi}(LL')\underline{u}(L') \tag{2.2}$$

using, e.g., periodic boundary conditions, one obtains the eigenvalues and eigenvectors of the system as the solutions $\omega_\lambda^2 \equiv \omega^2(\underline{q},j)$ and $\underline{e}_\lambda \equiv \underline{e}(\underline{q},j)$ of the secular equation

$$\det|\underline{D}(\underline{q}) - M\omega^2\underline{I}| = 0 \tag{2.3}$$

where the dynamical matrix $\underline{D}$ is defined as

$$\underline{D}(\kappa\kappa'|\underline{q}) = \sum_\ell \underline{\phi}(LL') \exp\{-i\underline{q}[\underline{x}(L) - \underline{x}(L')]\} . \tag{2.4}$$

6

$\underline{M} = (M_\kappa)$, i.e., it is a diagonal mass matrix, while $\underline{I}$ is a unit matrix, $\underline{I} = (\delta_{\alpha\beta}\delta_{\kappa\kappa'})$, where α,β are Cartesian indices. The index j labels the different branches belonging to every wave vector $\underline{q}$.

The most important entity to be known, in addition to the dispersion relation $\omega(\underline{q},j)$, is the phonon density of states

$$D(\omega) = \sum_j n_j(\omega) \tag{2.5}$$

with the contribution from branch j

$$n_j(\omega) = \sum_{\underline{q}} \delta[\omega - \omega(\underline{q}j)]$$

$$= V_a \int \frac{dS_{\underline{q}}}{|\nabla_{\underline{q}}\,\omega(\underline{q}j)|} \,, \tag{2.6}$$

where $dS_{\underline{q}}$ denotes a surface element in $\underline{q}$ space.

It should be noted that the frequencies do not uniquely determine the eigenvectors and force constants of a crystal [2.6]. Therefore, in a calculation of dynamical properties which contains phonon eigenvectors (such as Raman scattering) additional arguments from microscopic theory, etc., are required to demonstrate the physical significance of a set of (formal or model) force constants.

2.2 Rigid Ion Model (RIM)

The lattice energy of ionic crystals is mainly given by two-ion Coulomb interactions of pointlike charges. In diatomic cubic crystals with ionic charges, $Z_\kappa = \pm Ze$, this energy per lattice cell reads

$$U_{ion} = \frac{1}{2} \sum_{\ell,\kappa\kappa'}' \frac{Z_\kappa Z_{\kappa'} e^2}{|\underline{x}(\ell\kappa) - \underline{x}(0\kappa')|} = \alpha_M \frac{(Ze)^2}{r_0} \tag{2.7}$$

with the Madelung constant α_M and the nearest-neighbor distance r_0. At equilibrium, the resulting attractive forces are balanced by repulsive short-range electronic overlap forces. They are conveniently described by two-body central Born-Mayer potentials

$$V_{\kappa\kappa'}(r) = A_{\kappa\kappa'} \exp(-r/\rho_{\kappa\kappa'}) \,. \tag{2.8}$$

In this Born model the total lattice energy per cell is given by

$$U(r) = -\alpha_M \frac{(Ze)^2}{r} + \frac{1}{2} \sum_{\kappa\kappa'}' V_{\kappa\kappa'}(r) \,. \tag{2.9}$$

KELLERMANN used this model for a calculation of dispersion curves in alkali halides. Here, the force matrix can be split into a long-range Coulomb and a short-range repulsive part as

$$\underline{\phi} = \underline{\phi}^C + \underline{\phi}^R .$$ (2.10a)

The Coulomb force matrix is given by

$$\underline{\phi}^C = \underline{Z}\underline{C}\underline{Z} , \quad \underline{Z} = (Z_\kappa) ,$$ (2.10b)

with the "Coulomb matrix" $\underline{C}$. The Fourier transform of $\underline{\phi}$ [refer to the corresponding transform in (2.4)] is the rigid-ion dynamical matrix

$$\underline{D}^{RI}(\underline{q}) = \underline{Z}\underline{C}(\underline{q})\underline{Z} + \underline{R}(\underline{q}) .$$ (2.11)

$C(\underline{q})$ is the Fourier transform of $\underline{C}$ and has a nonanalytical part at $\underline{q} \rightarrow 0$ which leads to the splitting of the longitudinal-optic from the transverse-optic modes in the nonpolariton regime (Lyddane-Sachs-Teller relation). A specific technique originated by Ewald has been developed for the calculation of $\underline{C}(\underline{q})$ (refer to MARADUDIN et al. [2.7]). In the simplest nearest-neighbor approximation this "rigid ion" model reduces with the help of the equilibrium condition, $U'(r) = 0$, to a one-parameter model which gives a surprisingly good fit to the acoustic and the transverse optic modes. The calculated longitudinal optic mode frequency is, of course, too high since no electronic screening has been considered. In other words, $\varepsilon_\infty = 1$ in the RIM.

2.3 Dipole Models (DM)

The difficulties in describing properly the longitudinal optic modes and related entities may be overcome by an explicit consideration of the electronic charge distortion induced at a displaced ion (TOLPYGO, 1948; SZIGETI, 1948; see [2.8]). Several models have been developed during the last 20 years which treat this distortion in the dipole approximation (refer to [2.8-11]). The most pictorial of these models is the shell model.

2.3.1 Shell Model (SM)

This model, originally due to DICK and OVERHAUSER (1958) and COCHRAN (1959) (see [2.8]) introduces in addition to the ion vector displacement $\underline{u}(L)$ an electronic polarization coordinate $\underline{w}(L)$. The potential depends now in a bilinear form on all $\underline{w}(L)$ and $\underline{u}(L)$ of the lattice. The equations of motions are

$$M_\kappa \underline{\ddot{u}}(L) = - \frac{\partial \phi}{\partial \underline{u}(L)}$$ (2.12)

$$M_{el}\ddot{\underline{w}}(L) = -\frac{\partial\phi}{\partial\underline{w}(L)} = 0 \ . \tag{2.13}$$

The second equation is the adiabatic condition for the shell model. One now introduces Coulomb and short-range forces between electrons as well as between electrons and ions and proceeds similarly to the case of the RIM; one obtains then the dynamical matrix

$$\underline{D}(\underline{q}) = \underline{D}^{RI}(\underline{q}) + \underline{D}^{DIP}(\underline{q}) \ , \tag{2.14}$$

where $\underline{D}^{RI}$ is the dynamical matrix of the RIM and $\underline{D}^{DIP}$ that of the induced dipolar forces

$$\underline{D}^{DIP} = -(\underline{T} + \underline{Z}\underline{C}\underline{Y})(\underline{\bar{S}} + \underline{Y}\underline{C}\underline{Y})^{-1}(\underline{T}^+ + \underline{Y}\underline{C}\underline{Z}) \ . \tag{2.15}$$

The matrices $\underline{T}$ and $\underline{\bar{S}}$ represent the short-range electron-ion and electron-electron coupling. $\underline{Y}$ is an electronic charge matrix analogous to the ionic charge matrix $\underline{Z}$. The matrix $\underline{\bar{S}}$ consists of the short-range electron-electron coupling $\underline{S}$ and of the local shell-core coupling $\underline{K} = (K_\kappa)$. In the simplest version of the shell model (SSM) one assumes that short-range forces act through the shells only ($\underline{R} = \underline{T} = \underline{S}$) and, furthermore, one neglects the the polarizability of the positive ions as compared to that of the negative ions ($K_+ \to \infty$). In the nearest-neighbor approximation there are then two additional parameters, Y_- and K_-, as compared with Kellermann's RIM. They can be fitted to the optical constants in the following equations:

$$\mu\omega_{TO}^2 = R_o' - \frac{4\pi}{3}\frac{\varepsilon_\infty + 2}{3} Z_s^2 \frac{e^2}{V_a} \tag{2.16}$$

$$\mu\omega_{LO}^2 = R_o' + \frac{8\pi}{3}\frac{\varepsilon_\infty + 2}{3} Z_s^2 \frac{e^2}{V_a} \ . \tag{2.17}$$

Here μ is the reduced mass of the lattice cell. The effective Szigeti charge Z_s is given by

$$Z_s = Z_+ + \frac{Y_-}{1 + K_-/R_o} \ . \tag{2.18}$$

R_o' defines the "center" frequency

$$\omega_c^2 \equiv \frac{1}{3}(2\omega_{TO}^2 + \omega_{LO}^2) \tag{2.19}$$

$$\mu\omega_c^2 = R_o' = \frac{R_o}{1 + R_o/K_-} \ . \tag{2.20}$$

9

This three-parameter SM (A, Y_-, K_-) improves the agreement between experiment and theory remarkably as compared with the RIM (see, e.g., Fig.2.10 for NaI). A is the longitudinal force constant to nearest neighbors.

2.3.2 Extended Shell Model (ESM)

An alternative representation of the dynamical matrix, (2.14), has been suggested by COCHRAN [2.8]

$$D(\underline{q}) = \underline{R}' + \underline{Z}'\underline{C}'\underline{Z}' \tag{2.21}$$

with the effective entities

$$\underline{R}' = \underline{R} - \underline{T}\underline{S}^{-1}\underline{T}^+ \tag{2.22}$$

$$\underline{Z}' = \underline{Z} - \underline{T}\underline{S}^{-1}\underline{Y} \tag{2.23}$$

$$\underline{C}' = (\underline{I} + \underline{C}\underline{Y}\underline{S}^{-1}\underline{Y})^{-1}\underline{C} . \tag{2.24}$$

The general structure of (2.21) is completely analogous to that of the microscopic theory (refer to [2.11,12]).

This ESM is, of course, able to describe every type of phonon dispersion relation by introducing a sufficient number of parameters in the matrices $\underline{R}, \underline{S}, \underline{T}$. A satisfactory description of the data requires, however, the introduction of pseudo-polarizabilities with positive electronic shell charges Y_+ at the cation lattice sites [2.13].

2.3.3 Overlap Shell Model (OSM)

A real justification of this argument can be obtained by a discussion of the polarization effects in the overlap charge [2.14]. A particularly fruitful version of this type of shell model turned out to be its combination with valence force fields for the case of partially ionic tetrahedrally coordinated compounds [2.15] (refer to Sect. 2.3.9). A particularly appealing feature of the OSM is the drastic reduction of second nearest-neighbor forces which often allows a good description of data with only six parameters (A, B, Y_+, Y_-, K_+, K_-). A and B are nearest neighbor longitudinal and transversal coupling constants respectively.

2.3.4 Deformation Dipole Model (DDM)

The shell model, in its extended version, is equivalent to the deformation dipole model (DDM) originally developed by HARDY [2.9]. The equivalence can be shown if the DDM is completed with a polarization self-energy which is missing in the original version [2.16]. The inter-relation between both models has been illuminated in a recent microscopic calculation by ZEYHER [2.17].

2.3.5 Deformable Shell Model (DEFSM)

The concept of a dipolar electronic charge distortion in the dipole models may be generalized to models with general electronic charge distortions consistent with the local symmetry character of a phonon with respect to a particular "deformable" ion. For example, in a cubic diatomic lattice the radial coupling of an ion to its six nearest neighbors may be decomposed into three dipolar, one monopolar (breathing), and two quadrupolar deformabilities. We may define a generalized short-range dynamical matrix $\underline{R}'$ which in the case of cubic ionic crystals has the form

$$\underline{R}' = \underline{D}^{RI} + \underline{D}^{DEF} \tag{2.25}$$

with

$$\underline{D}^{DEF} = -\sum_\Gamma \underline{T}_\Gamma \underline{S}_\Gamma^{-1} \underline{T}_\Gamma^+ \tag{2.26}$$

where Γ denotes the different symmetry representations [i.e., the monopolar (Γ_1^+), dipolar (Γ_{15}^+), and quadrupolar (Γ_{12}^+) symmetry in the radial nearest-neighbor approximation] of the ionic deformabilities. The concept becomes useful in cases where a specific deformability resembles an important part of the excitation spectrum of an ion. The first attempt in the case of alkali halides was the introduction of the breathing shell model.

2.3.6 Breathing Shell Model (BSM) and Double Shell Model (DSM)

The Γ_1^+ deformability of the anions was introduced by SCHRÖDER [2.18] in order to improve the description of the longitudinal optic branches in particular near the L point in the Brillouin zone. While it seems that in these cases the introduction of dipolar pseudo-polarizabilities at the cation lattice sites may be preferable [2.14] (see Sect. 2.3.3), the idea of a "breathing" deformability may be appropriate in systems with a very localized dynamic charge transfer such as $Sm_xY_{1-x}S$. Furthermore, in superconducting crystals like Nb [2.19] and NbC [2.20], a combination of volume and quadrupole deformabilities (in the second case at the nonmetal lattice sites) seems to give a very satisfactory approach to the pronounced anomalies in the dispersion curves of these crystals.

An alternative approach in this case has been given by WEBER in his double-shell model (DSM) [2.21] which has been used with great success during the last few years. We note that many modifications of the BSM have been published under different names ("deformable shell model" DEFSM, etc.) (refer to BASU et al. [2.22]), but no new basic idea seems to originate with these proposals.

2.3.7 Quadrupole Shell Model (QSM)

A very interesting application of the quadrupolar deformability Γ_{12}^+ has been found in the case of silver halides by FISCHER et al. [2.23]. Here, the local virtual excitation of a d electron into an excited s state leads to a deformability of the right symmetry. Nearest-neighbor d-p hybridization may also contribute. The prediction by FISCHER et al. of an inversion of the transverse-mode eigenvectors in AgBr near the L point has been recently confirmed [2.24]. A microscopic calculation starting from the electronic band structure of silver halides has shown that the model is essentially correct [2.25]. Furthermore, a quadrupolar deformability may play a role in the lattice dynamics of copper halides and in the vibronic spectra of some oxides [2.26]. On the other hand, it seems doubtful whether the introduction of quadrupolar forces into the lattice dynamics of alkali halides is a useful concept [2.27].

2.3.8 Bond Charge Model (BCM)

In covalent crystals like diamond and its homologues the shell model and its modifications are useful models [2.28,29], although the electronic charge accumulation between the ions suggests a different description, i.e., the bond charge model (BCM). In its first version the bond charges (represented by point charges) had their positions always at the midpoint between neighboring ions [2.10]. In a recent extension by WEBER, the bond charge takes up the proper adiabatic force-free position in all displaced configurations. The two short-range forces are those of the KEATING model [2.31] which are essentially valence forces (see Sect. 2.3.9). A four-parameter version of the BCM gives a very satisfactory description of the dispersion curves of diamond, silicon, and germanium. The model, while sufficient for describing dispersion curves, fails in reproducing the complete crystal polarizability. Possibly the introduction of a charge transfer or a charge form factor could remove this failure.

2.3.9 Valence Force Field Model (VFFM)

This approach can be regarded as an extrapolation of the theory of normal vibrations of molecules. Here, the potential energy of the molecule is often given in terms of internal displacement coordinates, such as changes of bond distances, bond angles, etc. (refer to WILSON et al. [2.32]). The interatomic forces stem from distortions of the electronic charge density which is strongly concentrated along the inter-ionic connecting lines due to hybridization of the atomic orbitals. Since the covalent crystals behave in many respects like very large molecules, one should be able to describe them by an appropriate extension of molecular methods to almost infinite systems. The valence force field constants in these crystals may be obtained from an analysis of organic molecules with similar covalent bonds. For example, SCHACHTSCHNEIDER and SNYDER [2.33] have made a least-square determination of valence forces in branched paraffins which has been used successfully to determine the phonons in diamond, silicon, germanium, and gray tin [2.34].

It should be noted that the valence force field potential does not contain any Coulomb interactions which means that the interacting electronic and ionic charges are well screened. On the other hand, the VFFM lacks a clear relation to the dielectric properties of the crystals which gives the shell models their strong appeal.

2.3.10 Valence Overlap Shell Model (VOSM)

In the case of mixed ionic-covalent crystals a combination of the DDM with the VFFM has been used by KUNC et al. [2.35] for the zinc-blende structure compounds like ZnSe with fair success. A combination of the OSM with the VFFM (VOSM) describes quite successfully the lattice vibrations of many III-V, II-VI, and I-VII crystals [2.15]. In addition, this combination provides a useful basis for the analysis of second-order Raman spectra [2.15].

2.3.11 Three-Body Force Shell Model (TSM)

The deformabilities described in Sects. 2.3.4-2.3.7 are different combinations of three- and four-body forces. For example, the "breathing" deformability leads to a decrease of the bulk elasticity, $c_{11} + 2c_{12}$, while the quadrupolar (Γ_{12}^+) deformability decreases the shear elastic constant, $c_{11} - c_{12}$. Another possibility to introduce many-body forces is to start directly from the violation of the Cauchy relation, $c_{12} \neq c_{44}$ in cubic crystals, to determine many- (mainly three-) body forces (refer to [2.1,22]). This leads automatically to a type of deformable shell model but without a clear description in terms of local symmetry coordinates. SINGH and VERMA [2.36] introduced the so-called three-body-force shell model (TSM) which is, however, lacking a microscopic justification if $c_{12} - c_{44} < 0$ (refer to ZEYHER [2.17]). In this respect it is similar to the BSM if used for some alkali halides, but one should note that anharmonic corrections, which are very important for the Cauchy "violation", are not considered in those models. For recent refinements of the TSM see SINGH et al. in [2.38].

2.4 Microscopic Theory and Models

In this section a short comment is given on the present state of the art of the microscopic theory of phonons. For recent reviews the reader is referred to articles by SHAM and by BILZ et al. in [2.39] and to [2.1,10,12].

In microscopic theories one begins with the Coulomb electron-electron and electron-ion interactions and eventually derives the phonon dispersion relations by some approximation to the many-body Hamiltonian. The essential assumption is the definition of "ions" which include the tightly bound electrons. The Coulomb potential of the ions has then to be replaced by a pseudo-potential which takes account of the exchange and correlation effects of the valence electrons with the core electrons.

At the present time, two different approaches are used in microscopic theory. The first one describes the short-range inter-atomic forces as originating with the overlap of (mainly) nearest and (to a smaller extent) second-nearest neighbor charge densities. The overlap effect also leads to a modification of long-range Coulomb forces. A successful sequence of investigations [2.17] has recently led to the first parameter-free calculation of phonon dispersion curves in LiD [2.40].

The second approach is the dielectric function method. The basic ideas of this method have been developed in the 1960s (refer to SINHA [2.10]). During the last years the theory has concentrated on localized wave functions in order to obtain an explicit representation of the inverse dielectric function which appears in the dynamical matrix (refer to HANKE [2.12]). These treatments provide a microscopic basis for a generalized dipole model which may be applied to metals as well as to insulators and semiconductors.

14

2.5 Abbreviations Used in Figure Captions

$\omega(\underline{q})$	Phonon frequency as a function of wave vector
$\omega_c(\underline{q})$	Calculated phonon frequency only
$D(\omega)$	Density of states as a function of frequency
$D_m(\omega)$	Measured density of states (amplitude weighted)
[Ref. 7.8, Fig. 4]	Figure 4 in [7.8] at the end of this book
T = 300 K	Measured at 300 K temperature
P: 1.36 Kb	Measured at 1.36 Kb pressure
M: 7 P-FCM	Model used: seven parameter force constant model
3NN-MASM	3rd nearest neighbour modified axially symmetric force constant model
-FCM	Force constant model
-RIM	Rigid ion model
-SM	Shell model
-MSM	Modified shell model
-BSM	Breathing shell model
-DDM	Deformation dipole model
-TSM	Three-body force shell model
-DEFSM	Deformable shell model
-DBM	Deformable bond model
-BCM	Bond charge model
-VSM	Valence force shell model
-VDDM	Valence force deformation dipole model
-DSM	Double shell model
-ASM	Axially symmetric model
-MASM	Modified axially symmetric model
-VFFM	Valence force model
-OSM	Overlap shell model
-VOSM	Valence overlap shell model
-RLM	Rigid layer model
-QSM	Quadrupole shell model
M:-	No model used (guide lines to the eye)
Lit. [10.9]	For further information about this crystal see [10.9]

2.6 Conversion Factors Between Units Used to Express Phonon Energies

	1 THz	10^{13} rad s^{-1}	10^2 cm^{-1}	10 meV
1 THz	1.0000	0.6283	0.3336	0.4136
10^{13} rad s^{-1}	1.5916	1.0000	0.5309	0.6582
10^2 cm^{-1}	2.9979	1.8836	1.0000	1.2398
10 meV	2.4181	1.5193	0.8066	1.0000

Part II

Phonon Atlas of
Dispersion Curves and
Densities of States

3. Rare-Gas Crystals

Low-temperature measurements of all rare-gas crystals exist. The main interest is clearly in the different modifications of solid He. Here, the theory of quantum crystals has to be used [3.1,2], while the other rare-gas crystals are usually analyzed in terms of (modified) Lennard-Jones (6,12) potentials corresponding to 3NN-FCM.

The following systems are treated:

Crystal	Figures showing	
	Dispersion curve $\omega(\underline{q})$	Density of states $D(\omega)$
He^4 (fcc)	3.1	
He^4 (hcp)	3.2a	3.2b
He^4 (bcc)	3.3	
Ne	3.4a	3.4b
Ar	3.5a	3.5b
Kr	3.6a	3.6b
Xe	3.7a	3.7b

He⁴(fcc)

Fig. 3.1. He^4(fcc): ω($\underline{q}$) [Ref. 3.1, Fig. 4], T = 16 K, V$_m$ = 11,7 cm^3, dashed lines calc. [3.2]

Fig. 3.2a. He^4(hcp): $\omega(\underline{q})$ [Ref. 3.3, Fig. 4], T = 4.2 K, p = 0.23 kb, M: 18P-MASM, TA$_{\|}$, TO$_{\|}$ from Ref. 3.4

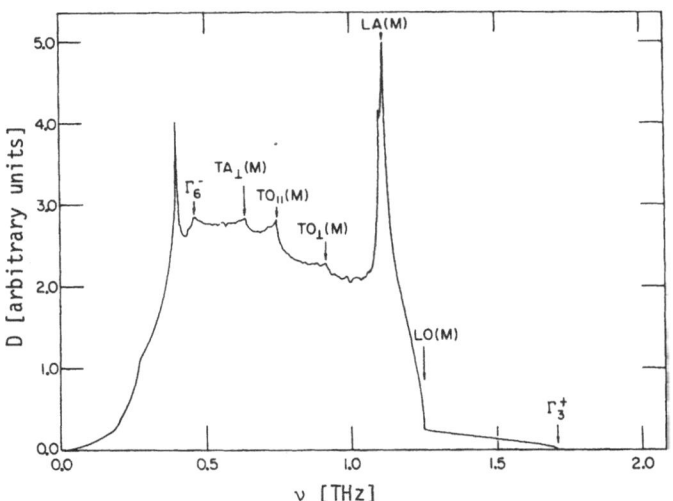

Fig. 3.2b. He^4(hcp): $D(\omega)$ [Ref. 3.3, Fig. 6], T = 4.2 K, p = 0.23 kb, M: 18P-MASM

He⁴(bcc)

Fig. 3.3. He⁴(bcc): $\omega(\underline{q})$ [Ref. 3.5, Fig. 7], T = 1.63 K, a = 4.12 Å, solid lines calc. [3.6,7]

Fig. 3.4a. Ne: ω(q) [Ref. 3.11, Fig. 7], T = 5 K, M: 3NN-FCM, Lit. [3.8,9,10]

Fig. 3.4b. Ne: D(ω)
[Ref. 3.11, Fig. 8],
T = 5 K, M: 3NN-FCM,
Lit. [3.8,9,10]

Ar

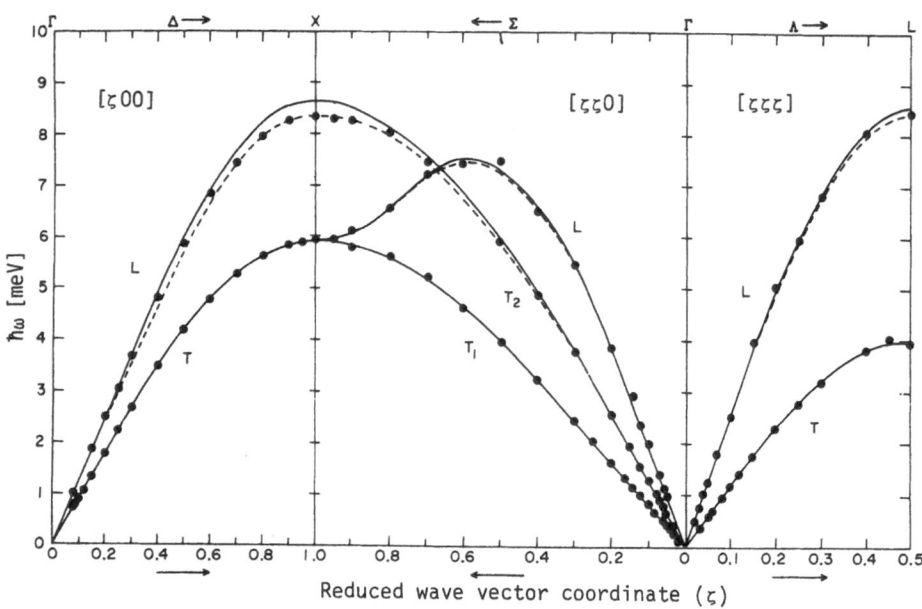

Fig. 3.5a. Ar: ω(q) [Ref. 3.12, Fig. 2], T = 10 K, M: 3NN-FCM (dashed lines), solid lines calc. [13]

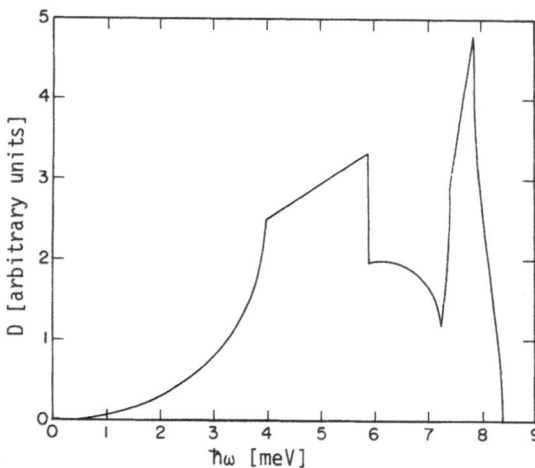

Fig. 3.5b. Ar: D(ω) [Ref. 3.12, Fig. 3], T = 10 K, M: 3NN-FCM

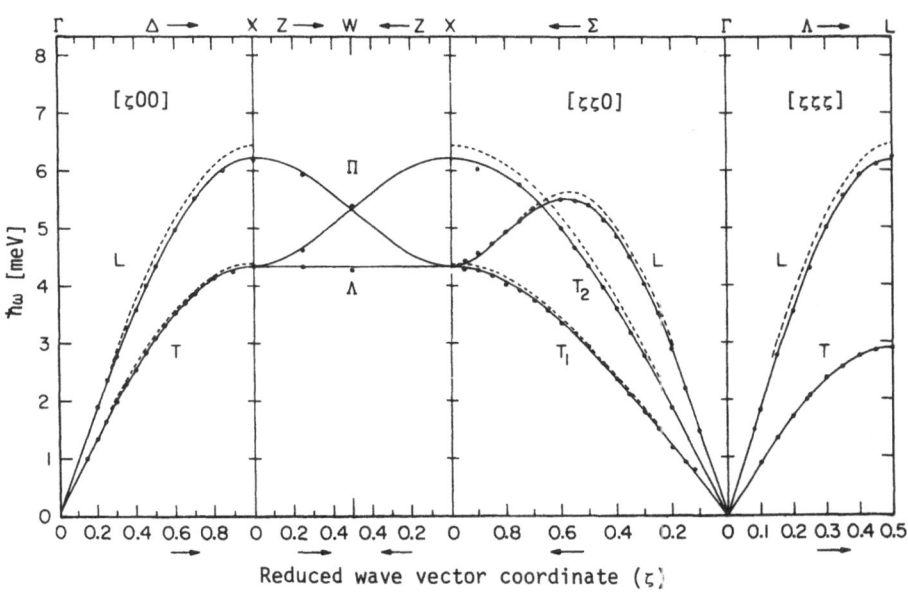

<u>Fig. 3.6a.</u> Kr: ω(q) [Ref. 3.14, Fig. 3], T = 10 K, M: 3NN-FCM (solid line), dashed line calc. [15]

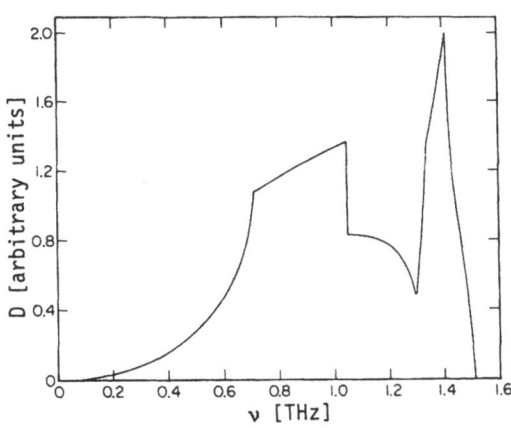

<u>Fig. 3.6b.</u> Kr: D(ω) [Ref. 3.14, Fig. 4], T = 10 K, M: 3NN-FCM

Xe

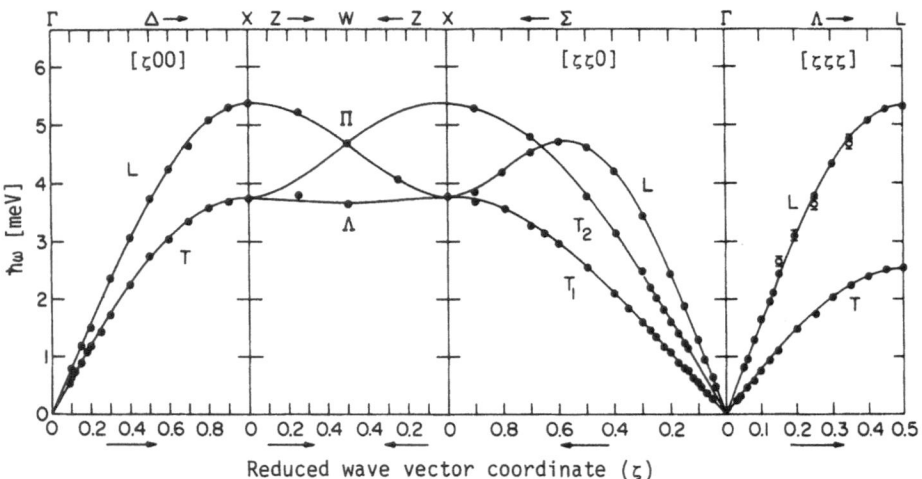

Fig. 3.7a. X: ω(q) [Ref. 3.16, Fig. 1], T = 10 K, M: 3NN-FCM,
open circles [3.17], T = 4 K

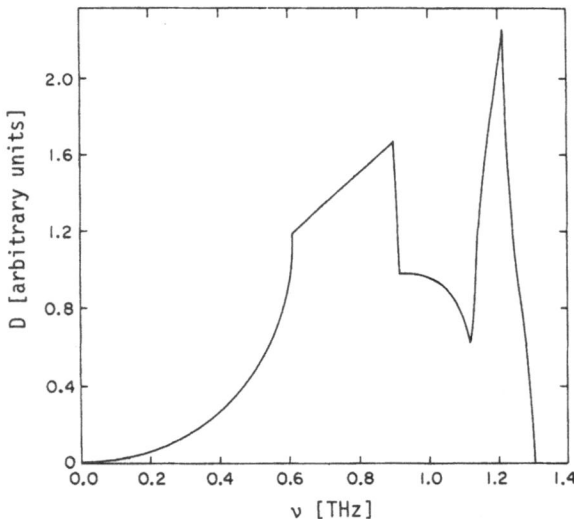

Fig. 3.7b. Xe: D(ω)
[Ref. 3.16, Fig. 2],
T = 10 K, M: 3NN-FCM

4. Alkali Halides (Rock Salt Structure)

Alkali halides are the best investigated family of crystals. Experiments are usually done at room temperature or at that of liquid nitrogen. Since the model theory (mainly dipolar models such as the SM or DM) as well as the microscopic theory [4.4] are highly developed (for recent reviews see [2.41, 4.8]) it seems to be worthwhile to measure phonons at helium temperature in a few alkali halides such as LiD, NaI, etc. This would provide a basis for a quantitative analysis of the anharmonic properties of these crystals which is still in its infancy.

The following systems are treated:

Crystal	Figures showing		Crystal	Figures showing	
	Dispersion curve $\omega(\underline{q})$	Density of states $D(\omega)$		Dispersion curve $\omega(\underline{q})$	Density of states $D(\omega)$
[7]LiH	4.1a	4.1c	NaI	4.10a	4.10b
[7]LiD	4.1b	4.1c	KF	4.11a	4.11b
LiF	4.2a	4.2b	KCl	4.12a	4.12b
LiCl	4.3		KBr	4.13a,b	4.13c
LiBr	4.4		KI	4.14a	4.14b
LiI	4.5		RbF	4.15a	4.15b
NaH		4.6	RbCl	4.16a	4.16b
NaF	4.7a	4.7b	RbBr	4.17a	4.17b
NaCl	4.8a	4.8b	RbI	4.18a	4.18b
NaBr	4.9a	4.9b			

⁷LiH ⁷LiD

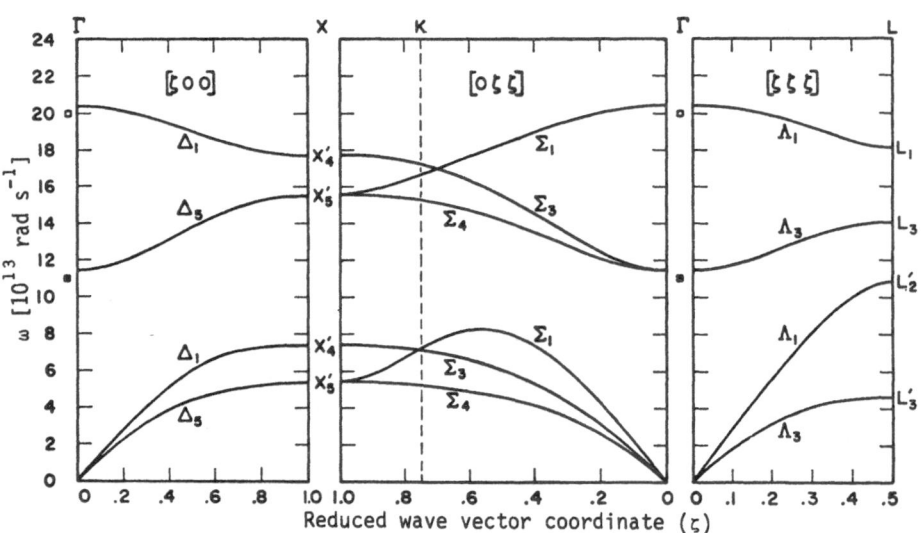

<u>Fig. 4.1a.</u> ^{7}LiH: $\omega_c(\underline{q})$ [Ref. 4.1, Fig. 4], T = 300 K, M: 7P-SM, Lit. [4.2]

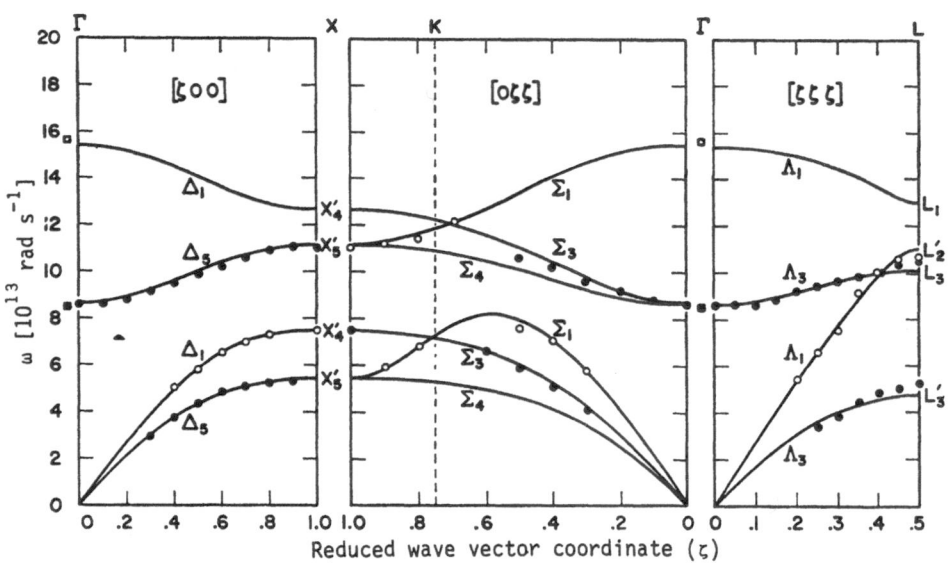

<u>Fig. 4.1b.</u> ^{7}LiD: $\omega(\underline{q})$ [Ref. 4.1, Fig. 3], T = 300 K, M: 7P-SM, Lit. [4.3-8]

28

Fig. 4.1c. ⁷LiH ⁷LiD: D(ω) [Ref. 4.1, Fig. 5], dashed lines: measured [4.9], solid lines: calculated curve, T = 300 K, M: 7P-SM

LiF

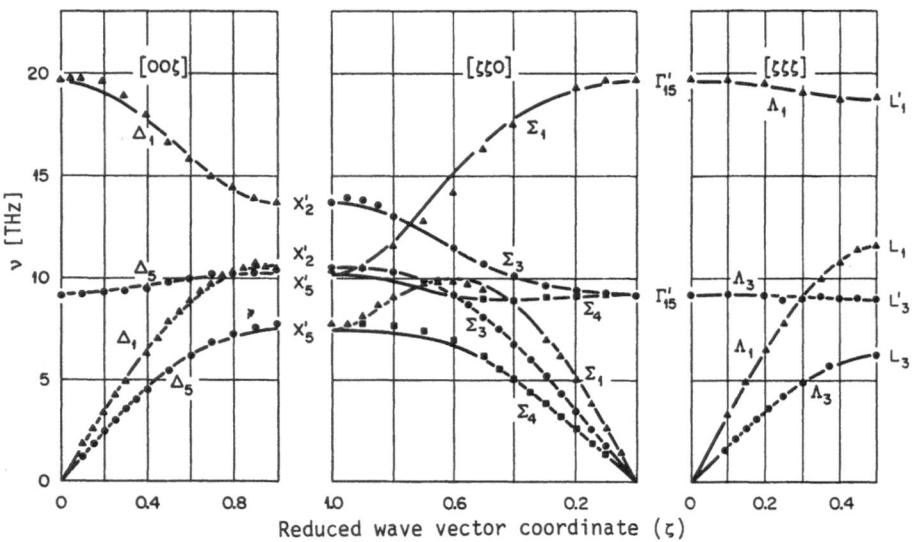

Fig. 4.2a. LiF: ω(q) [Ref.4.10, Fig.4], T = 298 K, M: 7P-SM,
Lit. [4.2,11-13,64]

Fig. 4.2b. LiF: D(ω) [Ref.4.10, Fig.6a], T = 298 K, M: 7P-SM, Lit.[4.2]

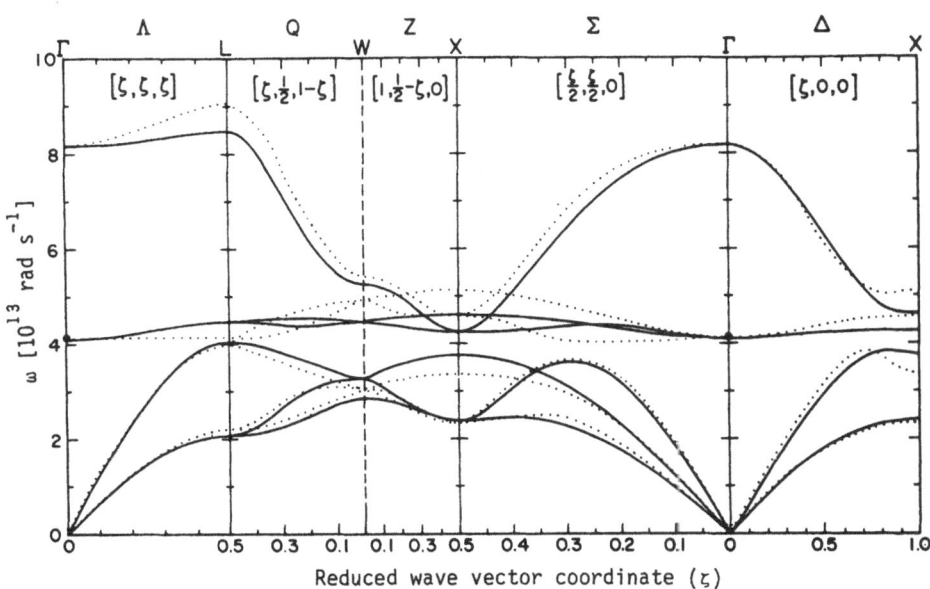

Fig. 4.3. LiCl: $\omega_c(\underline{q})$ [Ref. 4.13, Fig. 1], T = 0 K, M: 7P-BSM (full lines), 7P-DDM (dotted lines), Lit. [4.2]

LiBr

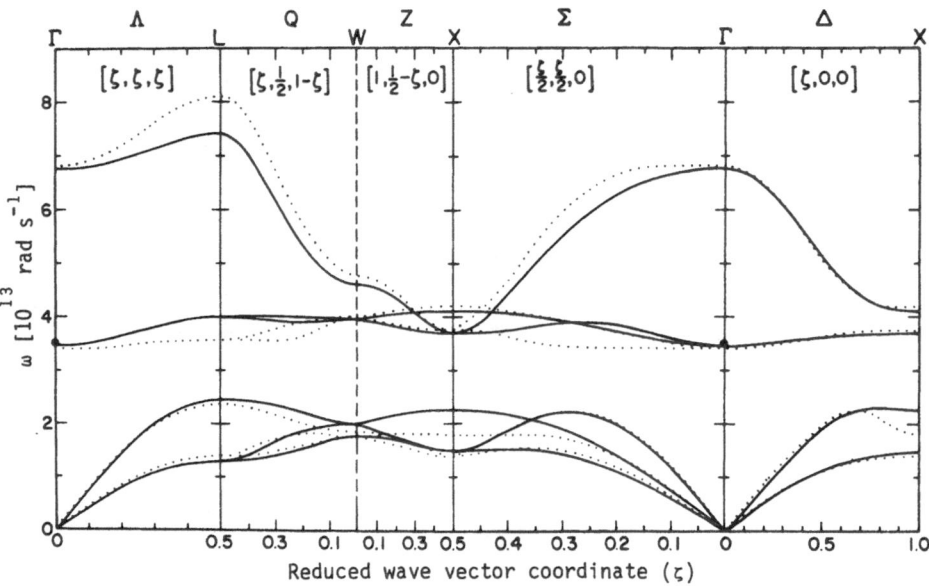

Fig. 4.4. LiBr: $\omega_c(\underline{q})$ [Ref. 4.13, Fig. 1], T = 0 K, M: 7P-BSM (full lines), 7P-DDM (dotted lines), Lit. [4.14]

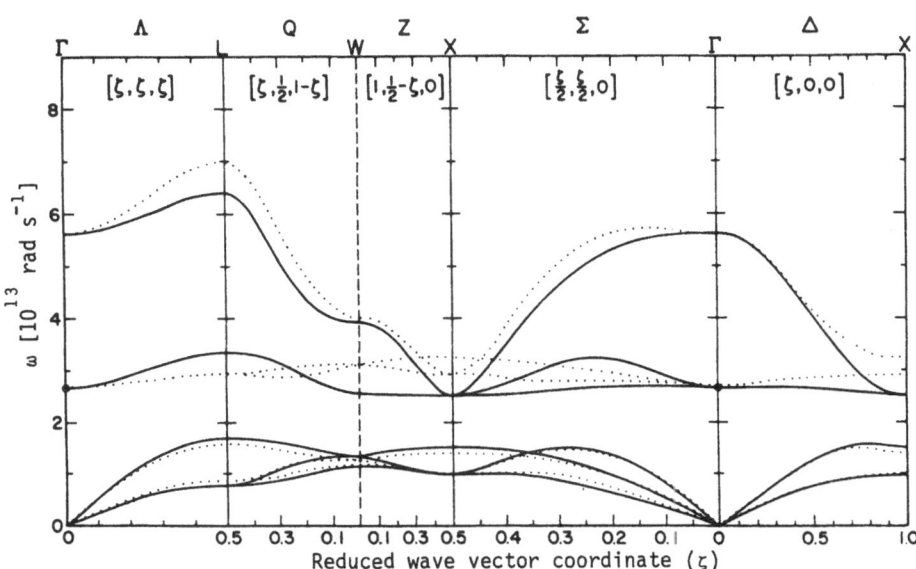

Fig. 4.5. LiI: $\omega_c(\underline{q})$ [Ref. 4.13, Fig. 1], T = 290 K, M: 7P-BSM (full lines), 7P-DDM (dotted lines), Lit. [4.14]

Fig. 4.6. NaH: D(ω) [Ref. 4.15, Fig. 7], T = 90 K

Fig. 4.7a. NaF: ω(q) [Ref.4.16, Fig.5], T = 295, M: 9P-SM (full lines),
5P-SM (dashed lines), Lit. [4.17-24]

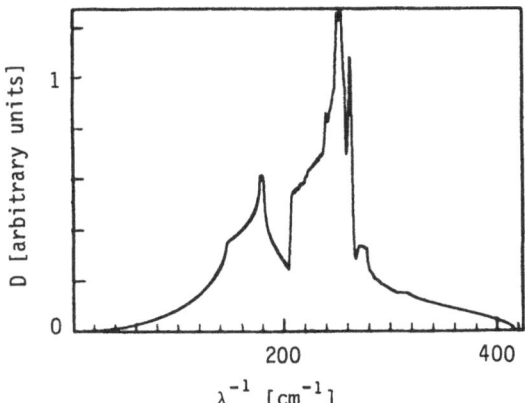

Fig. 4.7b. NaF: D(ω)
[Ref. 4.25, Fig 4],
M: 11P-SM, Lit.[4.19,20]

NaCl

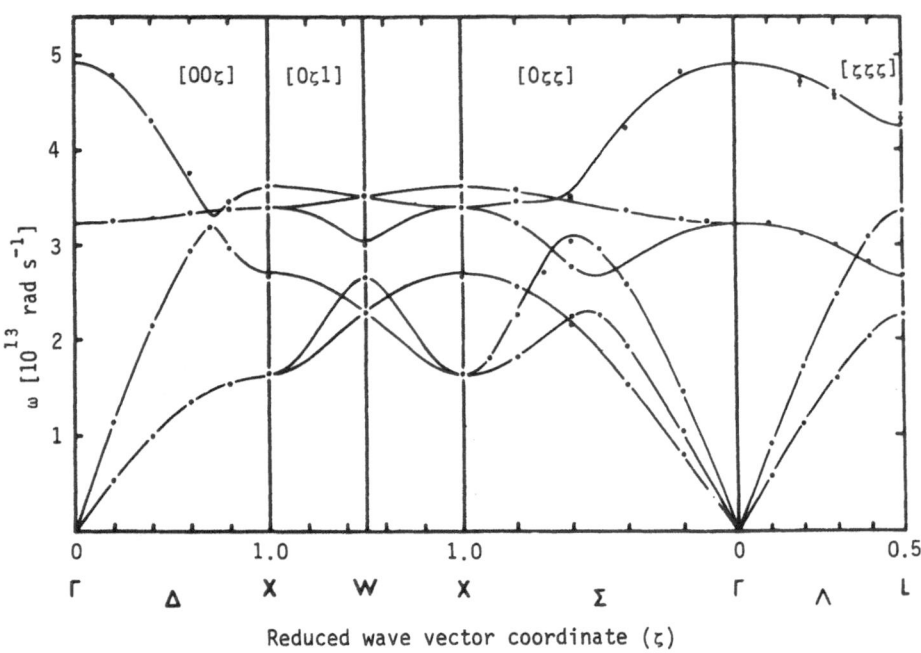

Fig. 4.8a. NaCl: ω(q) [Ref. 4.26, Fig. 1a], T = 80 K, M: 11P-SM,
Lit. [4.11,18,20,21,24,25,27-33]

Fig. 4.8b. NaCl: D(ω)
[Ref. 4.25, Fig. 4],
T = 80 K, M: 11P-SM,
Lit. [4.26]

△

Fig. 4.9a. NaBr: ω(q) [Ref. 4.34, Fig. 7], T = 295 K, M: 7P-SM (solid lines), 9P-SM (dashed lines), 8P-BSM (dotted lines), Lit. [4.18,20,21,24,25,35,63]

Fig. 4.9b. NaBr: D(ω) [Ref. 4.34, Fig. 8], M: 9P-SM

NaI

Fig. 4.10a. NaI: ω(q) [Ref. 4.36, Fig. 1], T = 100 K, Measurements [4.37,38],
M: 7P-BSM (full lines), 7P-SM (dashed lines), 5P-FCM (dotted lines),
Lit. [4.18,20,24,36-43]

Fig. 4.10b. NaI: D(ω)
[Ref. 4.44, Fig. 4b],
M: 9P-SM

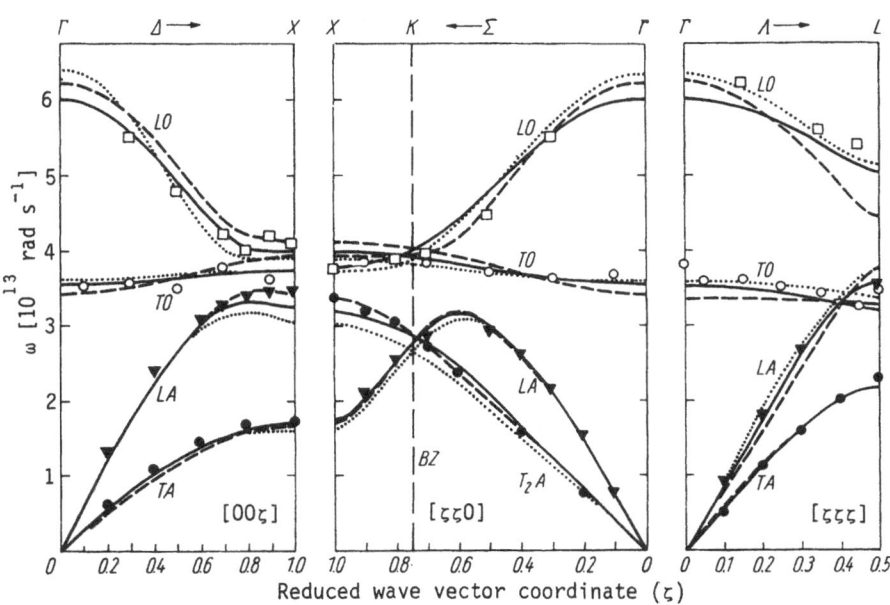

Fig. 4.11a. KF: ω(q) [Ref. 4.45, Fig. 1], T = 300 K, M: 11P-SM (solid-lines), 7P-FCM (dashed lines), 9P-TSM (dotted lines), Lit. [4.46]

Fig. 4.11b. KF: D(ω) [Ref. 4.45, Fig. 2], M: 11P-SM

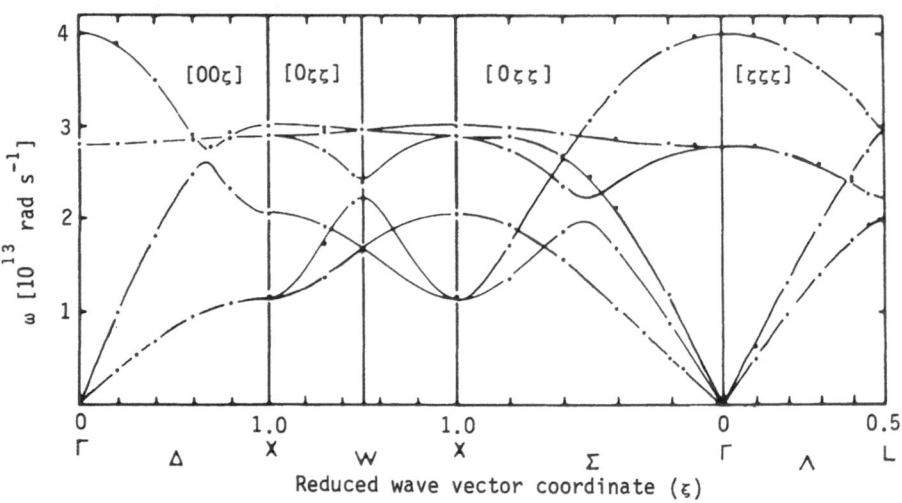

Fig. 4.12a. KCl: ω(q) [Ref. 4.26, Fig. 1b], T = 80 K, M: 11P-SM, Lit. [4.30, 46-50,64]

Fig. 4.12b. KCl: D(ω) [Ref. 4.26, Fig. 2b], T = 80 K, M: 11P-SM, Lit. [4.25]

Fig. 4.13a. KBr: ω(q) [Ref. 4.36, Fig. 2], T = 90 K, Measurements [4.37],
M: 7P-BSM (solid-lines), 7P-SM (dashed lines), 5P-FCM (dotted lines),
Lit. [4.25,38,49,51-54]

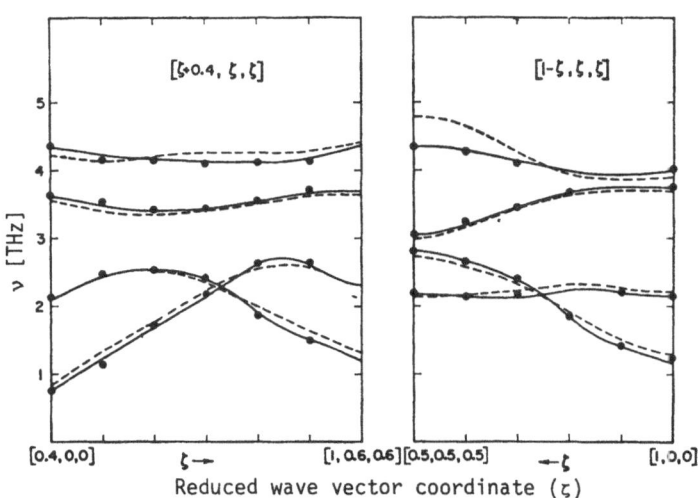

Fig. 4.13b. KBr: ω(q) [Ref. 4.51, Fig. 3], T = 90 K, M: 9P-SM (solid lines),
4P-SM (dashed lines), Lit. [4.25,38,39,49,51-54]

KBr

Fig. 4.13c. KBr: D(ω) [Ref. 4.25, Fig. 4], M: 11P-SM, Lit. [4.51]

Fig. 4.14a. KI: ω(q) [Ref. 4.55, Fig. 1], T = 95 K, M: 11P-SM (solid lines),
3P-SM (dashed lines), Lit. [4.25,35,54]

Fig. 4.14b. KI: D(ω) [Ref. 4.56, Fig. 9], T = 95 K, M: 11P-SM, Lit. [4.25]

RbF

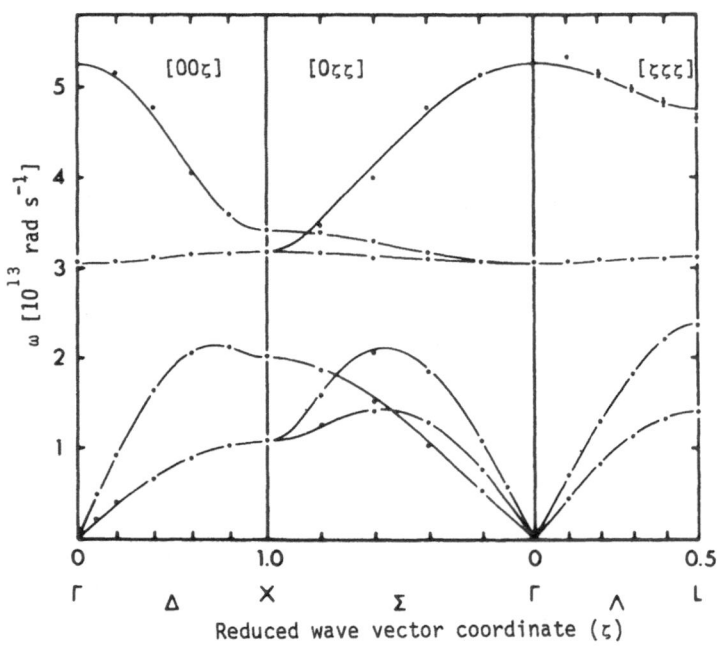

Fig. 4.15a. RbF: $\omega(\underline{q})$ [Ref. 4.26, Fig. 1d], T = 80 K, M: 11P-SM,
Lit. [4.57,58,65]

Fig. 4.15b. RbF: $D(\omega)$ [Ref. 4.26,
Fig. 2d], T = 80 K, M: 11P-SM

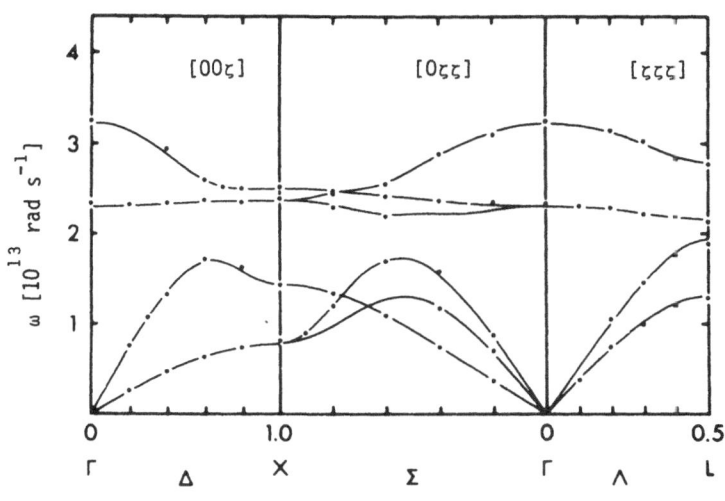

Reduced wave vector coordinate (ζ)

<u>Fig. 4.16a.</u> RbCl: $\omega(\underline{q})$
[Ref. 4.26, Fig. 1c], T = 80 K,
M: 11P-SM, Lit. [4.57-59,65]

<u>Fig. 4.16b.</u> RbCl: $D(\omega)$ [Ref. 4.26,
Fig. 2c], T = 80 K, M: 11P-SM

RbBr

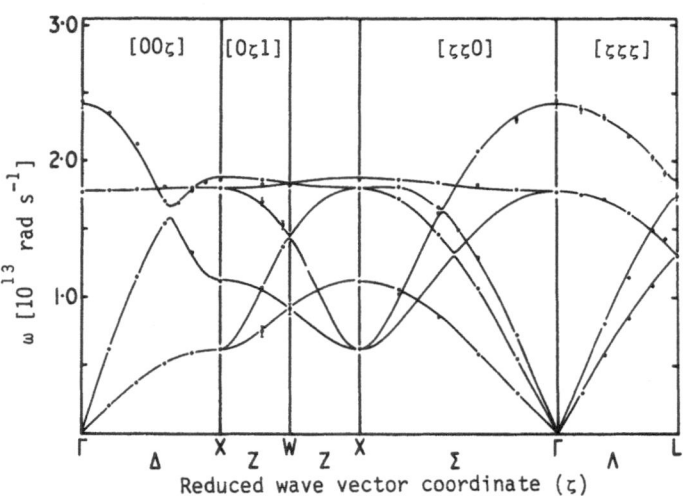

<u>Fig. 4.17a.</u> RbBr: ω($\underline{q}$) [Ref.4.60, Fig.1], T = 80 K, M: 11P-SM, Lit.[4.58,65]

<u>Fig. 4.17b.</u> RbBr: D(ω) [Ref.4.60, Fig.2], T = 80 K, M: 11P-SM

Fig. 4.18a. RbI: ω(q) [Ref.4.61, Fig.1], T = 80 K, M: 11P-SM, Lit. [4.58,65]

Fig. 4.18b. RbI: D(ω) [Ref.4.61, Fig.2], T = 80 K, M: 11P-SM

5. Metal Oxides (Rock Salt Structure)

These crystals show a behavior very similar to that of alkali halides. Shell models with about ten parameters give generally good agreement with experimental data. Of particular interest is CoO because of its phase transition from the antiferromagnetic to the paramagnetic state.

The following systems are treated:

Crystal	Figures showing	
	Dispersion curve $\omega(\underline{q})$	Density of states $D(\omega)$
MgO	5.1a	5.1b
CaO	5.2a	5.2b
SrO	5.3a	5.3b
BaO	5.4a	5.4b
MnO	5.5	
FeO	5.6a	5.6b
CoO (paramagnetic)	5.7a	
CoO (antiferromagnetic)	5.7b	
NiO	5.8a	5.8b

MgO

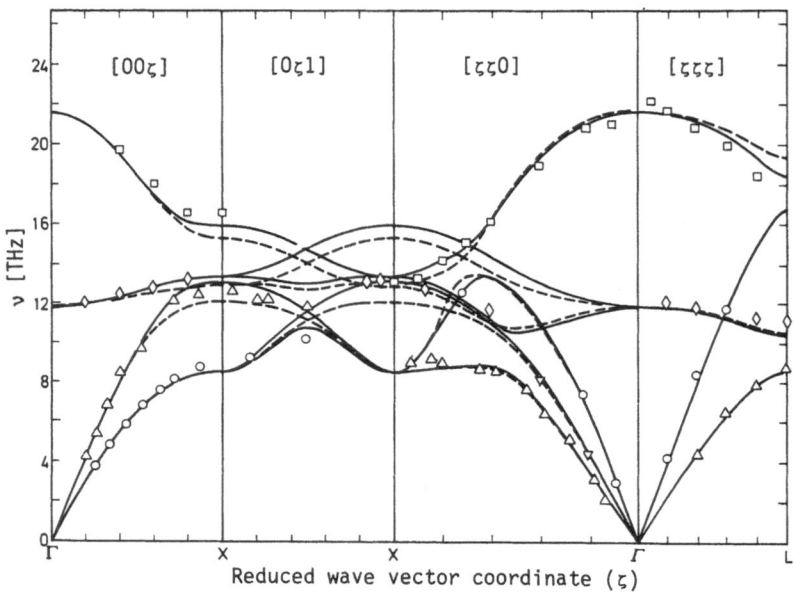

<u>Fig. 5.1a.</u> MgO: ω(q) [Ref. 5.1, Fig. 2], T = 293 K, M: 8P-BSM (full lines), 7P-SM (dashed lines), Lit. [5.2-7,5.25]

<u>Fig. 5.1b.</u> MgO: D(ω) [Ref. 5.1, Fig. 3], M: 8P-BSM, Lit. [5.26]

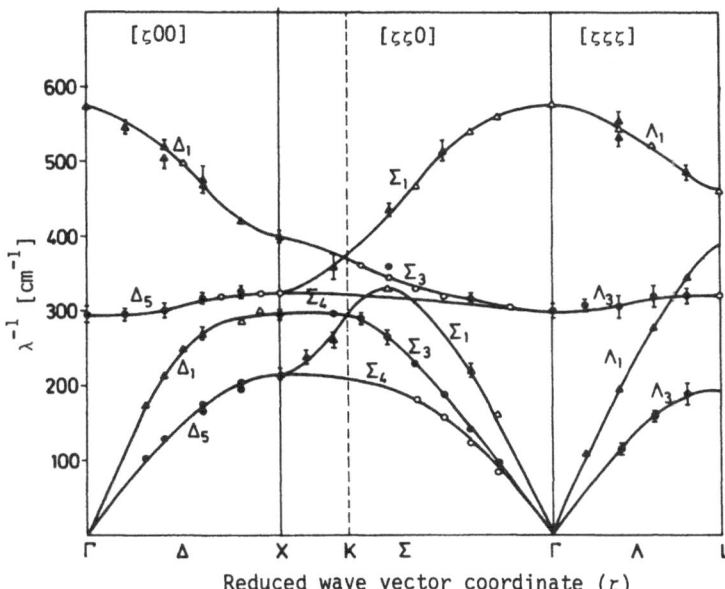

Fig. 5.2a. CaO: ω(q) [Ref. 5.7, Fig. 2], T = 293 K, full circles and triangles
[5.8], open circles and triangles [5.9], M: 11P-SM, Lit. [5.10]

Fig. 5.2b. CaO: D(ω) [Ref. 5.8, Fig. 2], T = 293 K, M: 11P-SM, Lit. [5.9]

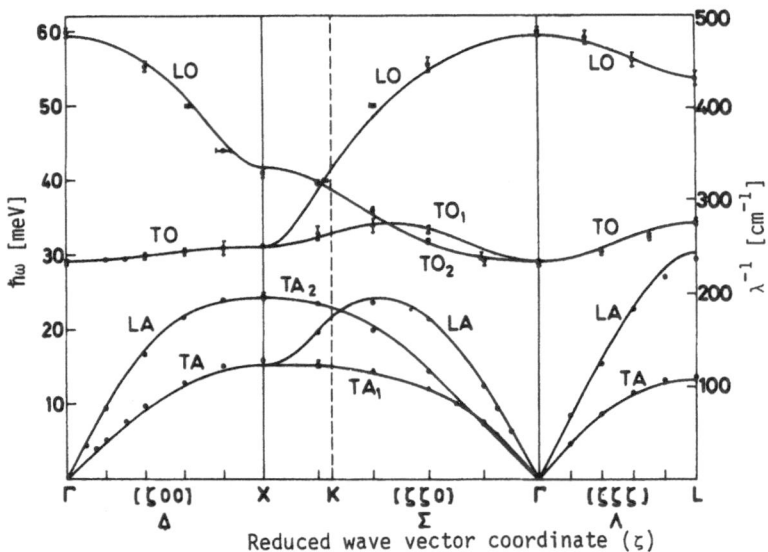

Fig. 5.3a. SrO: ω(q) [Ref. 5.11, Fig.1], T = 300K, M: 7P-SM, Lit.[5.10,12]

Fig. 5.3b. SrO: D(ω) [Ref. 5.11, Fig. 2], T = 300 K, M: 7P-SM

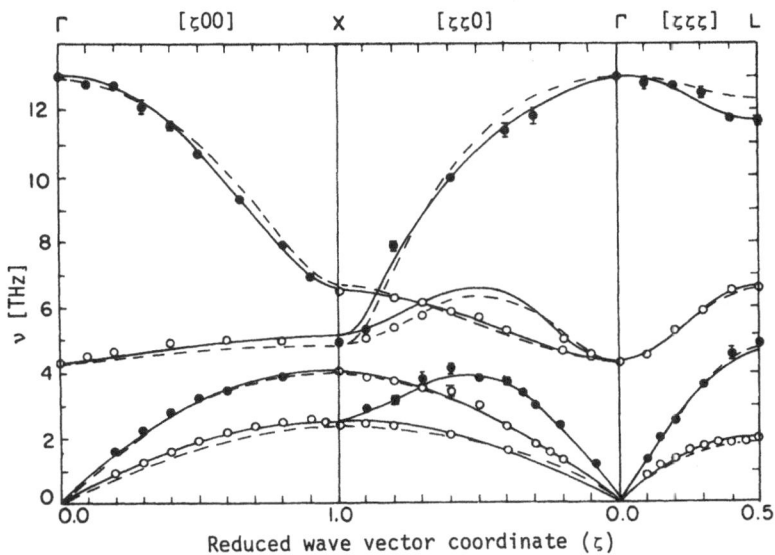

<u>Fig. 5.4a.</u> BaO: ω(q) [Ref. 5.13, Fig. 1], T = RT, M: 15P-SM (full lines), 8P-TSM (dashed lines)

<u>Fig. 5.4b.</u> BaO: D(ω) [Ref. 5.13, Fig. 2], T = RT, M: 15P-SM

MnO

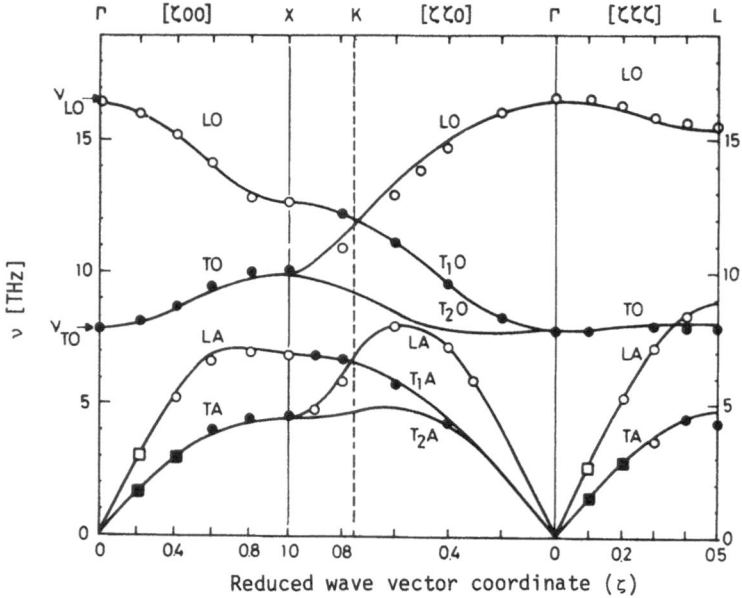

Fig. 5.5. MnO: ω(q) [Ref. 5.17, Fig. 1], T = RT, M: 9P-OSM, Lit. [5.15-18]

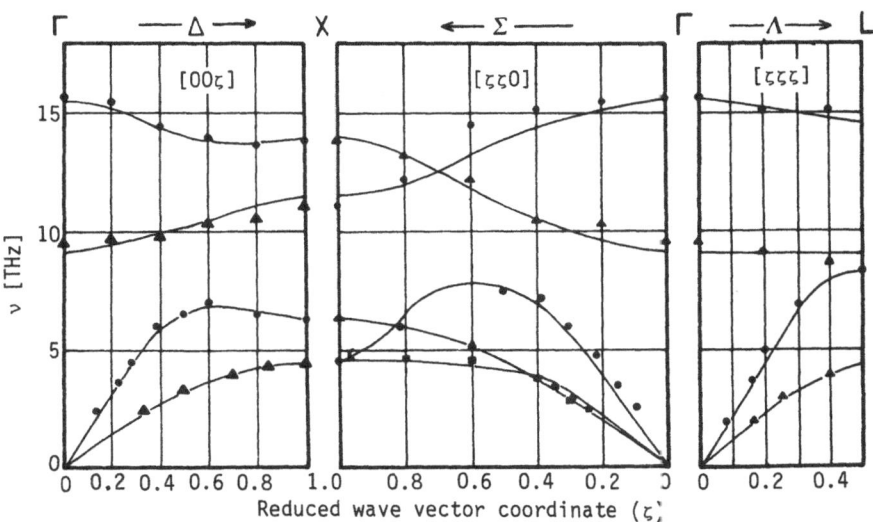

Fig. 5.6a. FeO: $\omega(q)$ [Ref. 5.19, Fig. 5], T = RT, M: 12P-SM, Lit. [5.14]

Fig. 5.6b. FeO: $D(\omega)$ [Ref. 5.19, Fig. 9], M: 12P-SM, two different channel widths

Fig. 5.7a. CoO (paramagnetic): ω(q) [Ref. 5.20, Fig. 5], T = 4.25 K for [00ξ] and [ξξ0], T = 330 K for [ξξξ], M: 11P-SM, Lit. [5.18,21]

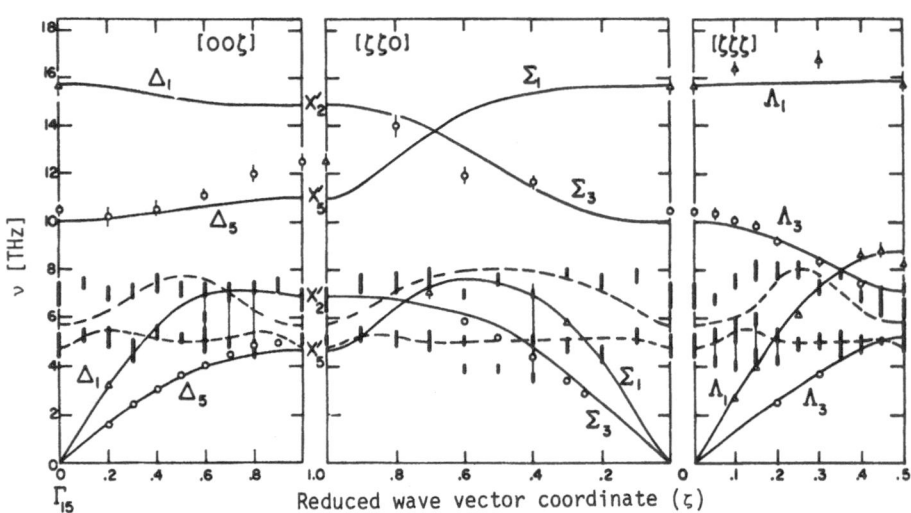

Fig. 5.7b. CoO (antiferromagnetic): ω(q) [Ref. 5.20, Fig. 4], T = 110 K M: 11P-SM, Lit. [5.18,21]

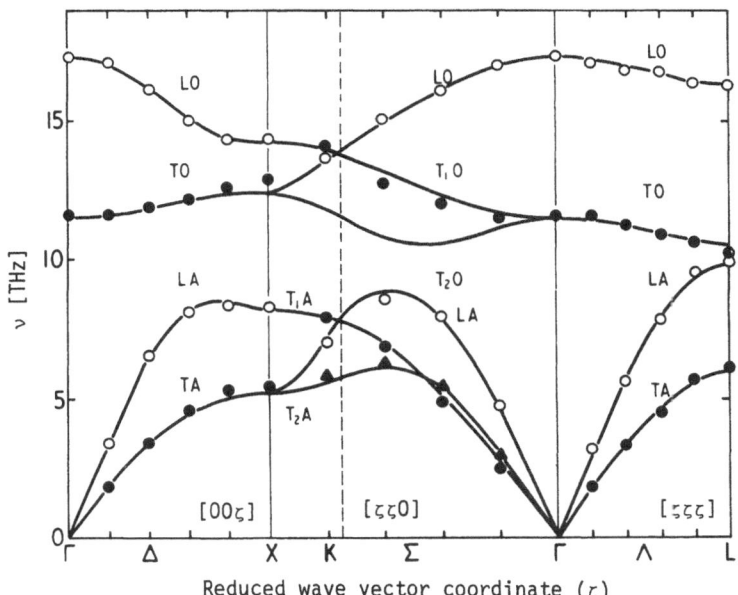

Fig. 5.8a. NiO: ω(q) [Ref. 5.22, Fig. 2], T = 293 K, M: 9P-SM, Lit. [5.14,18,23,24]

Fig. 5.8b. NiO: D(ω) [Ref. 5.23, Fig. 2], T = 297 K, M: 11P-SM

6. Transition Metal Compounds (Rock Salt Structure)

The interest in these groups of materials is related to the fact that many of them (the group V_b carbides and the group IV_b nitrides) are high-temperature superconductors. The general behavior of the phonon dispersion curves is still similar to those in other rock salt crystals and may well be described in terms of a shell model. Characteristic deviations appear in the optic modes at long wavelengths where the metallic screening destroys the Lyddane-Sachs-Teller splitting between the LO and TO modes, while at higher q values the Coulomb splitting between these modes can be clearly recognized. While this behavior is typical for all metallic crystals with optic modes, the phonon anomalies at q vectors with components q_α = 1/2 (in particular LA (1/2,0,0) and TA,LA at the L point (1/2,1/2,1/2) seem to correlate with high T_c's in a way typical for this family of compounds. A successful model description was given by WEBER in his DSM [6.6,13], while different microscopic explanations were given in [6.19,25].

Crystal	Figures showing		Crystal	Figures showing	
	Dispersion curve $\omega(\underline{q})$	Density of states $D(\omega)$		Dispersion curve $\omega(\underline{q})$	Density of states $D(\omega)$
YS	6.1a	6.1b	NbC_{1-x}	6.8a,b	6.8c
TiC	6.2a	6.2b	NbN (γ-)	6.9a	
TiN	6.3a	6.3b	NbN (δ-)		6.9b
ZrC	6.4a	6.4b	TaC	6.10	
ZrN		6.5	UC	6.11	
HfC	6.6		UN	6.12	
VC	6.7a	6.7b			

Note added in proof:

The phonon dispersion curves of ZrN have recently been measured [A.N. Christensen, O.W. Dietrich, W. Kress, W.D. Teuchert: Phys. Rev. B *19*, 5699 (1979)]

Fig. 6.1a. YS: $\omega(\underline{q})$ [Ref. 6.1, Fig. 1], T = RT, M: 11P-DSM (full lines), 5P-SM (dashed lines), Lit. [6.2]

Fig. 6.1b. YS: $D(\omega)$ [Ref. 6.2, Fig. 2], T = RT, M: 11P-DSM (full lines), guide lines to the eye (dashed lines)

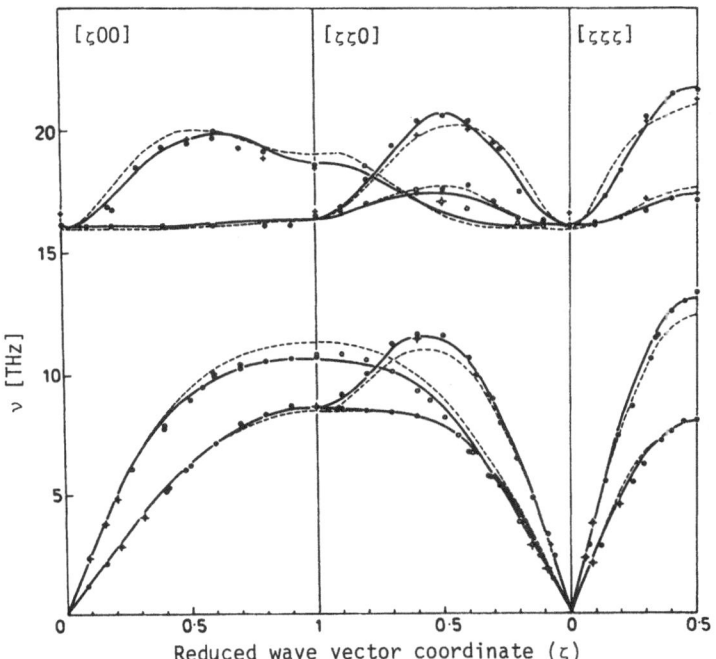

<u>Fig. 6.2a.</u> TiC: ω(q) [Ref. 6.3, Fig. 1], T = RT, circles TiC$_{0.95}$, crosses TiC$_{0.89}$, M: 12P-SM (full lines), 8P-SM (dashed lines)

<u>Fig. 6.2b.</u>
TiC$_{0.95}$: D(ω)
[Ref. 6.3, Fig. 2]
T = RT, M: 12P-SM

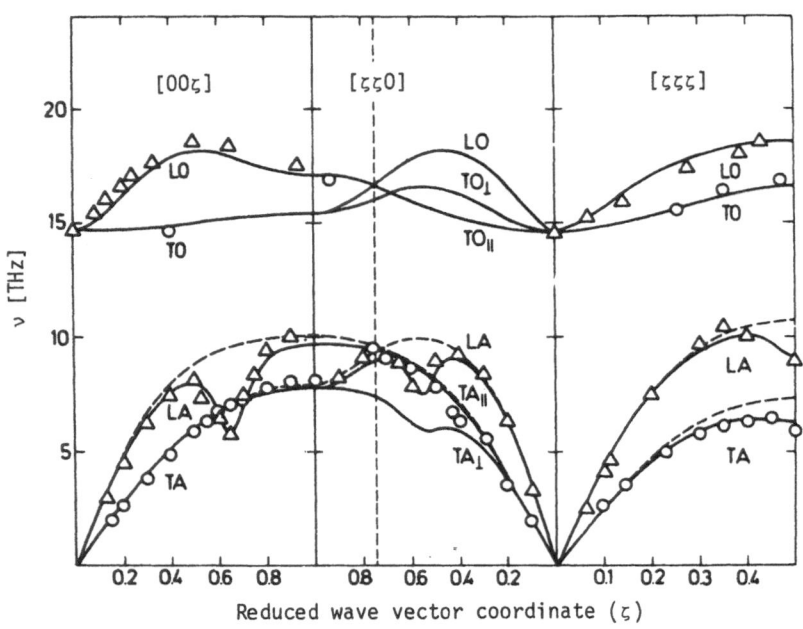

<u>Fig. 6.3a.</u> TiN$_{0.98}$: $\omega(\underline{q})$ [Ref. 6.4, Fig. 1], T = RT, M: 13P-DSM (full lines), 7P-SM (dashed lines)

<u>Fig. 6.3b.</u> TiN: D(ω) [Ref. 6.5, Fig. 2a], T = 296 K, M: -, bulk (full lines), microcrystalline (dashed lines), Lit. [6.4]

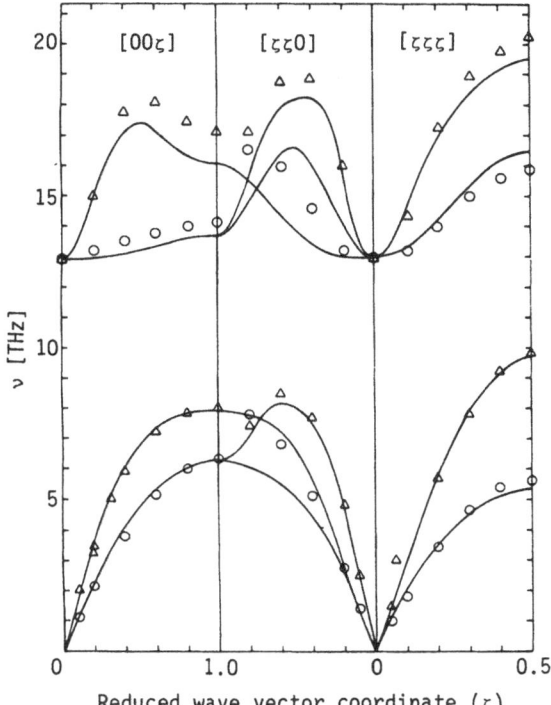

Fig. 6.4a. ZrC: ω(q̲)
[Re⁼. 6.6, Fig. 6],
T = 298 K, M: 9P-SM,
measurements [6.7,8],
Lit. [6.9]

Fig. 6.4b. ZrC: D(ω) [Ref. 6.10, Fig. 13], T = RT

Fig. 6.5. ZrN: D(ω) [Ref. 6.11, Fig. 1], T = RT

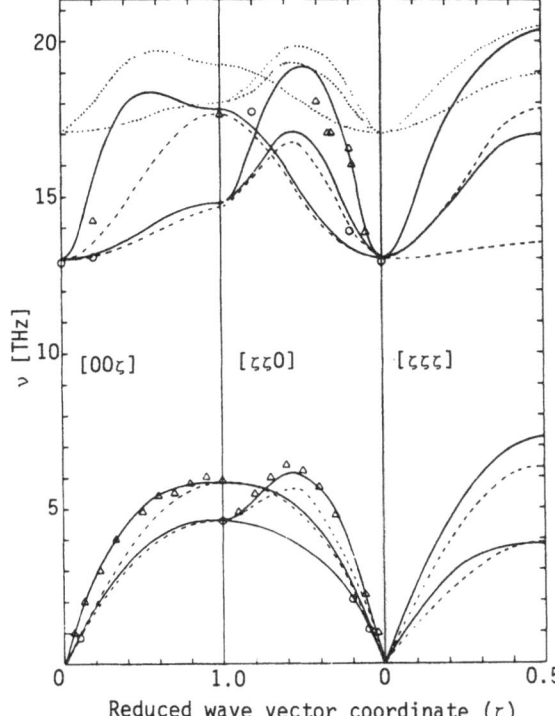

ν [THz]

[00ζ]   [ζζ0]   [ζζζ]

Reduced wave vector coordinate (ζ)

Fig. 6.6. HfC: ω(q)
[Re⁼. 6.6, Fig. 5],
T = 298 K, M: 9P-SM
measurements [6.12]
Lit. [6.13-15]

Fig. 6.7a. $VC_{0.87}$: $\omega(\underline{q})$ [Ref. 6.16, Fig. 1], T = RT, M: 9P-SM + 2P free electron screening

Fig. 6.7b. $VC_{0.87}$: $D(\omega)$ [Ref. 6.26, Fig. 1], T = RT

Fig. 6.8a. NbC$_{1-x}$
$\omega(\underline{q})$ [Ref. 6.17,
Fig. 3],
T = 298 K, M: -,
Lit. [6.6,18-21,25]

Fig. 6.8b. NbC$_{0.89}$:
$\omega(\underline{q})$ [Ref. 6.18,
Fig. 5], T = 298 K,
data for NbC$_{0.98}$
[6.17], M: 14P-DSM
(full lines),
dashed lines: width
of neutron groups,
Lit. [6.6,17,19-21,25]

ħω [meV]

Fig. 6.8c. NbC: D(ω) [Ref. 6.6, Fig. 10], M: 14P-DSM, acoustic phonons only

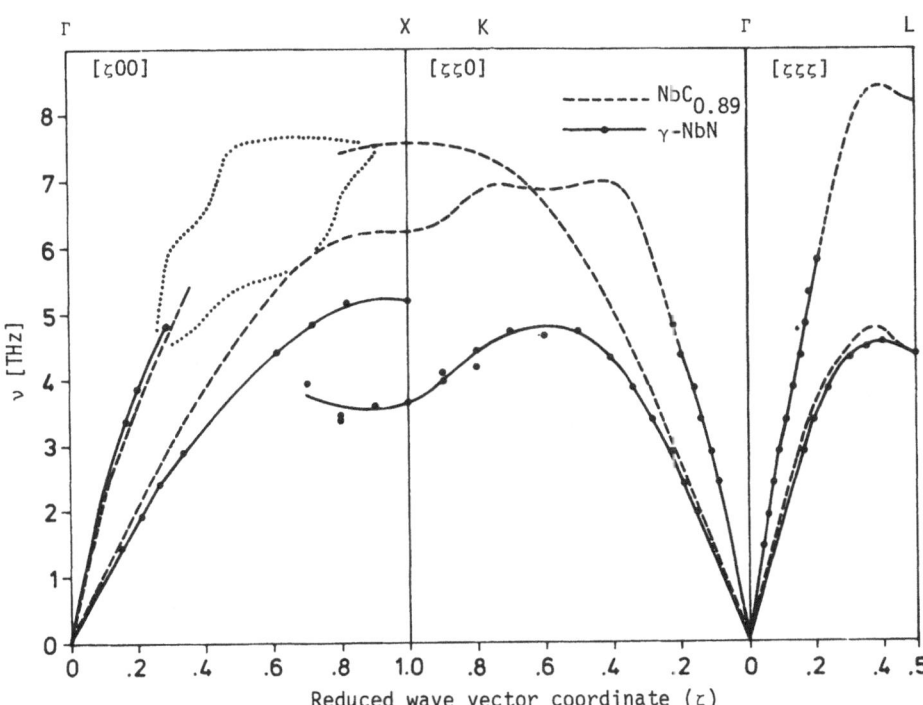

Fig. 6.9a. γ-NbN: ω(q̲) [Ref. 6.18, Fig. 1], M: -, Lit. [6.6]

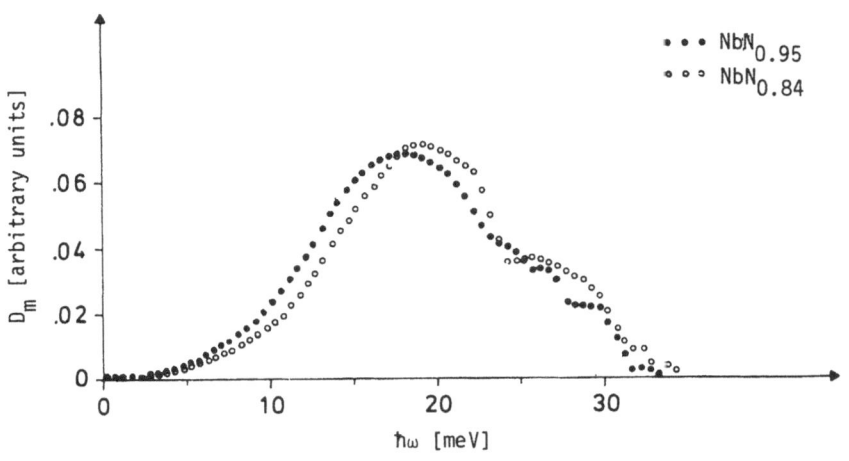

Fig. 6.9b. δ-NbN: D(ω) [Ref. 6.22, Fig. 1], M: -, Lit. [6.6]

TaC

Fig. 6.10. TaC: ω(q) [Ref. 6.17, Fig. 1], T = 298 K, 4.2 K, M: -,
Lit. [6.6,12-15]

<u>Fig. 6.11.</u> UC: ω(q) [Ref. 6.7, Fig. 6], T = 298 K, M: 2NN-FCM,
Lit. [6.6,14,23]

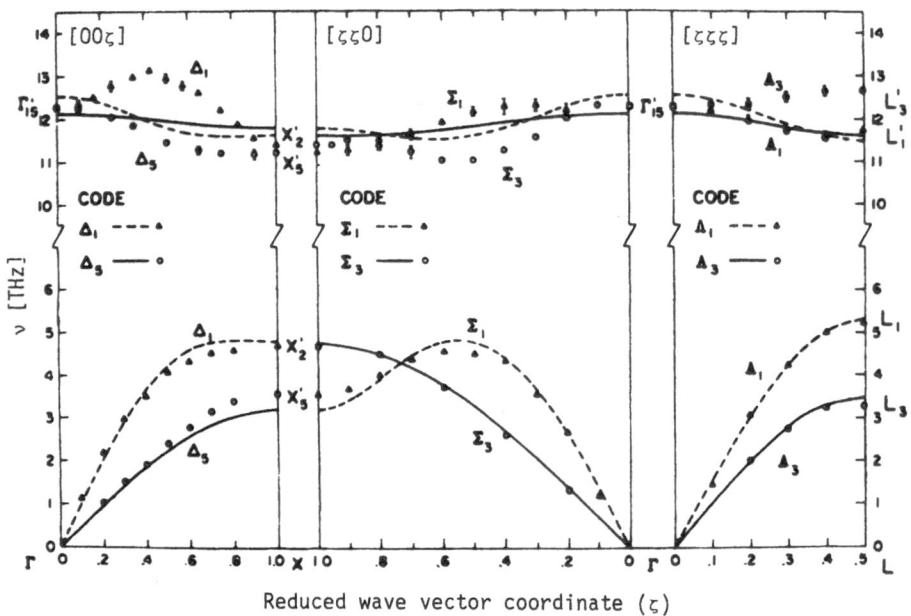

Fig. 6.12. UN: ω(q) [Ref. 6.24, Fig. 1], T = 4.2 K, M: 7P-RIM

7. Other Cubic Crystals (Rock Salt Structure)

This is physically the most interesting family of crystals among the "simple" cubic crystals. The first group (SnTe, etc.) exhibits interesting dielectric features due to the high polarizability of cations and anions. The silver halides show a characteristic deviation from other rock salt halides in the TO branch near all q points with components $q_\alpha \sim 1/2$. The origin of these anomalies is the quadrupolar deformability of the silver ion due to virtual electronic $d \rightarrow s$ excitations. The alkali cyanides behave like molecular crystals with (strongly temperature-dependent) rotational excitations and phase transitions.

The following systems are treated:

Crystal	Figures showing	
	Dispersion curve $\omega(\underline{q})$	Density of states $D(\omega)$
SnTe	7.1a	7.1b
PbS	7.2a	7.2b
PbSe	7.3a	7.3b
PbTe	7.4a	7.4b
NdSb	7.5a	7.5b
AgCl	7.6a	7.6b
AgBr	7.7a	7.7b
NaCN	7.8	
KCN	7.9a	7.9b
ND_4I	7.10a	7.10b
CsF	7.11a	7.11b

Note added in proof: The dispersion curves $\omega(\underline{q})$ of the SmS and $Sm_{0.75}Y_{0.25}S$ are found in [7.37] and [7.38] respectively.

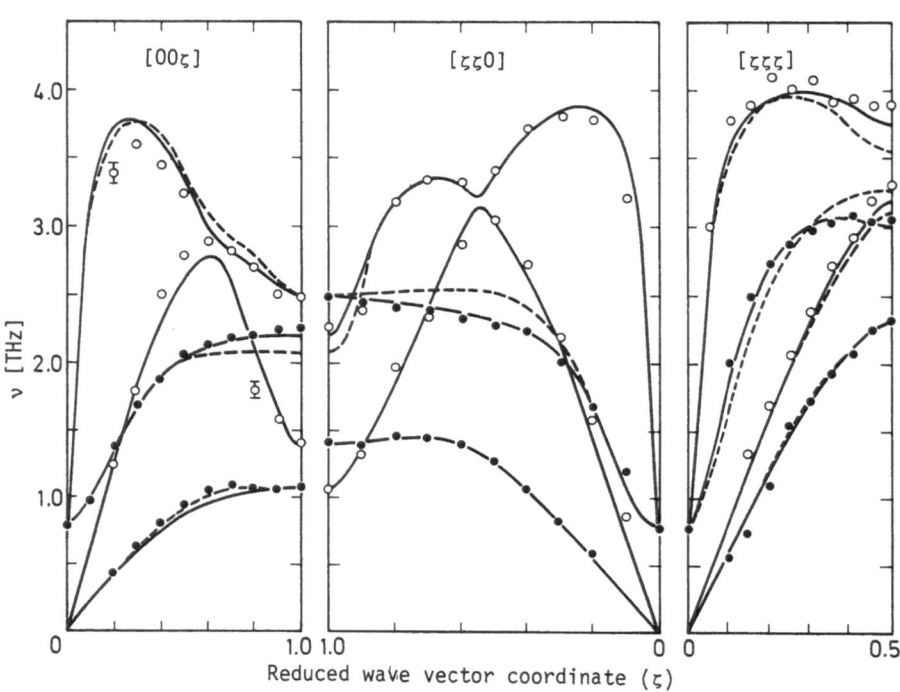

Fig. 7.1a. SnTe: ω(q) [Ref. 7.1, Fig. 1], T = 100 K, M: 15P-SM, full lines
Z ≠ 0, dashed lines Z = 0, Lit.[7.2]

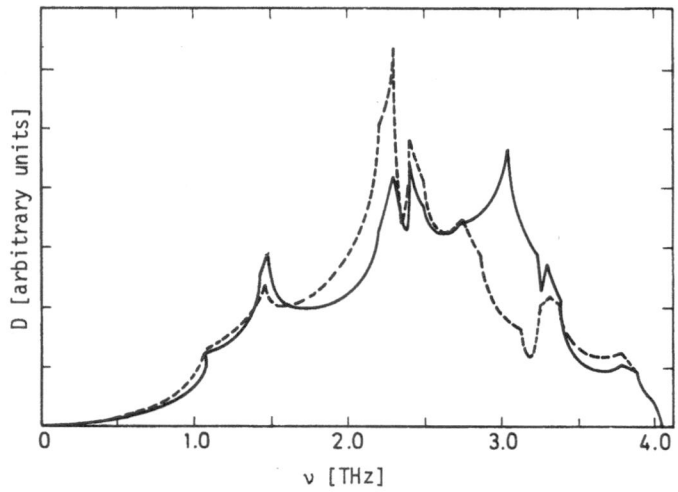

Fig. 7.1b. SnTe: D(ω)
[Ref.7.1, Fig.2],
full lines; dashed
lines: diagonal
element of the Green
function, Lit.[7.2]

<u>Fig. 7.2a.</u> PbS: ω(q̲) [Ref. 7.3, Fig. 1], T = 296 K, M: 14P-SM (solid lines),
7P-FCM (dashed lines)

<u>Fig. 7.2b.</u> PbS: D(ω) [Ref. 7.3, Fig. 2], M: 14P-SM

PbSe

Fig. 7.3a. PbSe: ω(q) [Ref. 7.4, Fig. 1], T = 296 K, M: 7P-SM, Lit. [7.3,35]

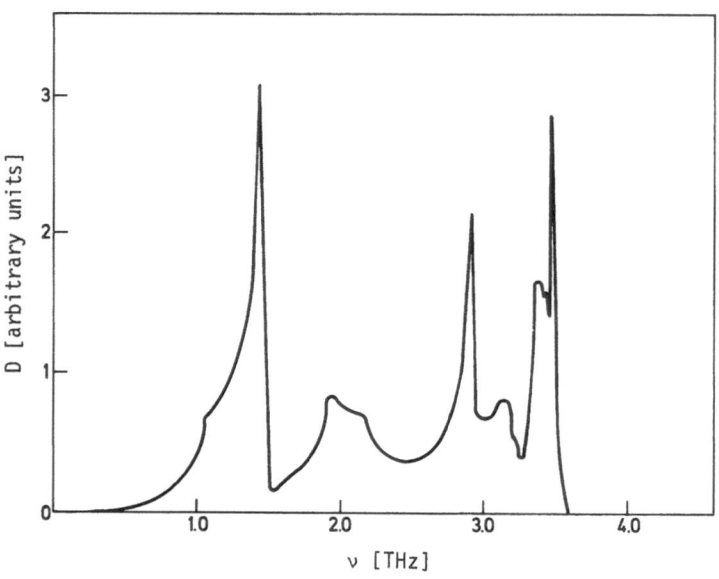

Fig. 7.3b. PbSe: D(ω) [Ref. 7.36], T = 296 K, M: 13P-SM, Lit. [7.3,35]

Fig. 7.4a. PbTe: ω(q̲) [Ref. 7.5, Fig. 1], T = 296 K, M: 13P-SM (full lines), 7P-FCM (dashed lines), Lit. [7.6]

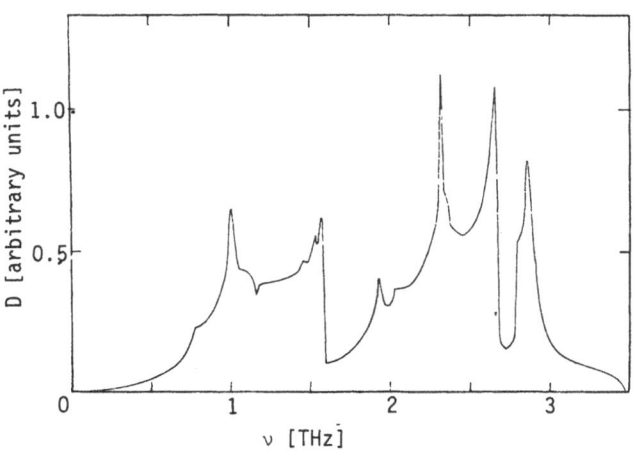

Fig. 7.4b. PbTe: D(ω) [Ref. 7.5, Fig. 2], M: 13P-SM (full lines)

NdSb

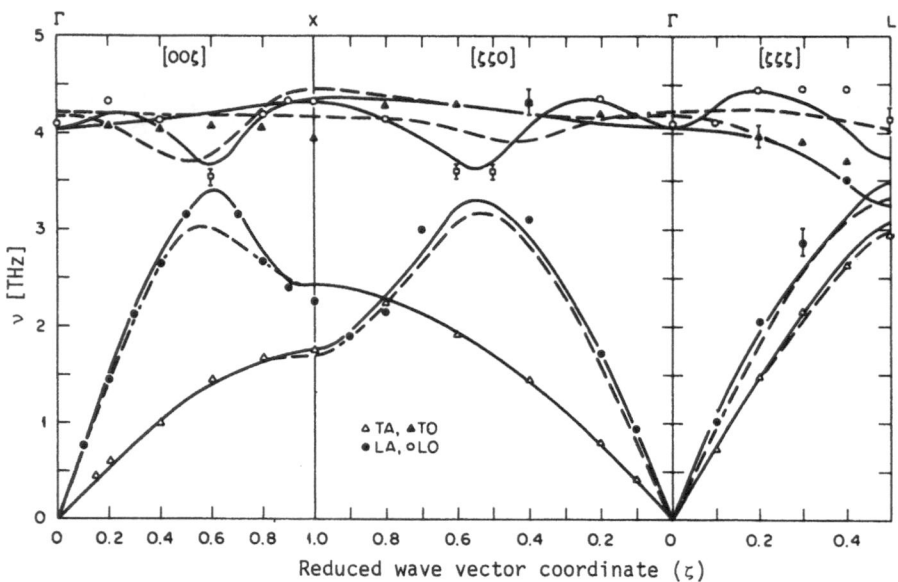

Fig. 7.5a. NdSb: $\omega(\underline{q})$ [Ref. 7.7, Fig. 1], T = RT, M: 9P- screened RIM (full lines), 2NN-SM (dashed lines)

Fig. 7.5b. NdSb: $D(\omega)$ [Ref. 7.7, Fig. 2], M: 9P- screened RIM

<u>Fig. 7.6a.</u> AgCl: ω(q) [Ref. 7.8, Fig. 1], T = 78 K, M: 13P-SM, Lit. [7.9-11]

<u>Fig. 7.6b.</u> AgCl: D(ω)
[Ref. 7.8, Fig. 3],
T = 78 K, M: 13P-SM

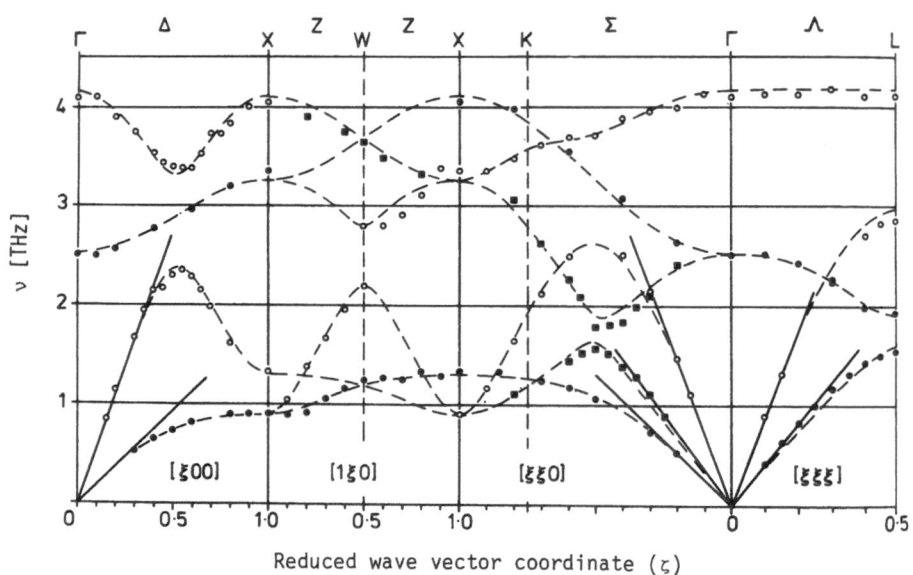

Fig. 7.7a. AgBr: ω(q) [Ref. 7.14, Fig. 1], T = 85 K, M: 15P-QSM,
Lit. [7.10,12-20]

Fig. 7.7b. AgBr: D(ω) [Ref. 7.14, Fig. 4], T = 85 K, M: 15P-QSM, total
density (full lines), partial densities of individual ions (dashed lines),
Lit. [7.10,16-18]

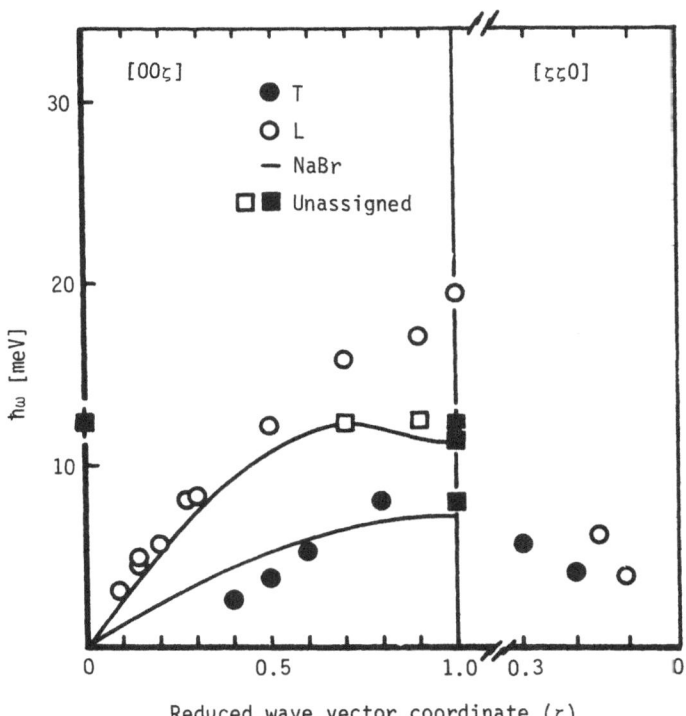

Fig. 7.8. NaCN: ω(q) [Ref. 7.21, Fig. 2], T = RT, (solid lines: dispersion curves of NaBr), Lit. [7.22,23]

△

Fig. 7.9a. KCN: ω(q) [Ref. 7.24, Fig. 1],
T = 215 K: ○ TA, □ T$_2$A, △ LA; T = 175 K:
⊙ TA, ⊡ T$_2$A, △ LA, LO; ●, ▲, □ overlapping
symbols, + [7.21], x [7.25], ◇ [7.26], full
lines: 7P-SM+2P dumb-bell model [7.22],
Lit. [7.21,23,27], further experimental data
are given in [7.34]

• 5K
• 77K
• 145K
• 180K

Fig. 7.9b. KCN: D(ω) [Ref. 7.27, Fig. 1],
T = 5 K, 77 K, 145 K, 180 K

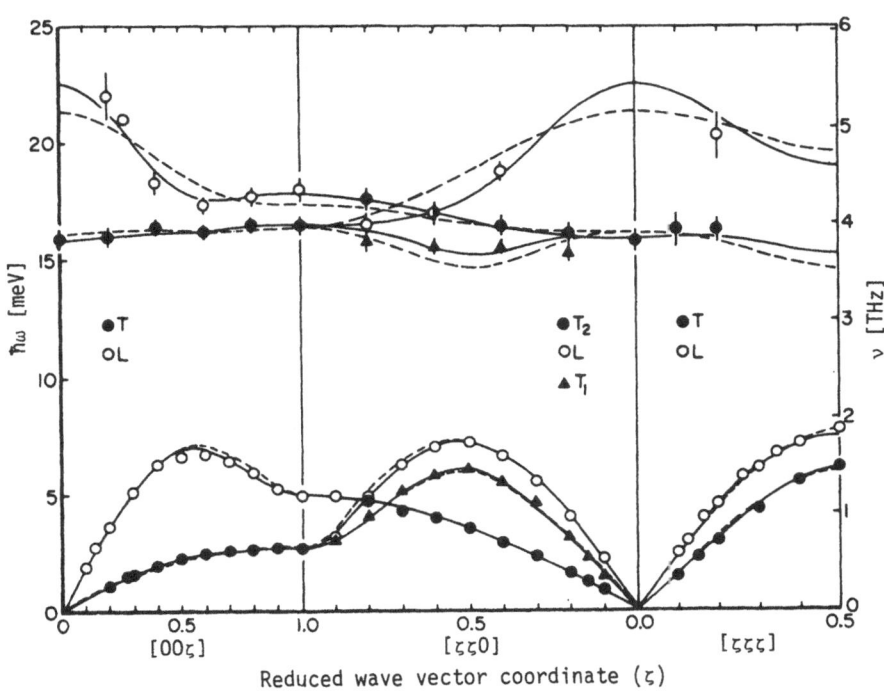

Fig. 7.10a. ND$_4$I: ω(q) [Ref.7.28, Fig.1], T = 296 K, M: 16P-BSM (full lines), 16P-SM (dashed lines), Lit. [7.29]

Fig. 7.10b. ND$_4$I: D(ω) [Ref. 7.28, Fig. 3], M: 16P-BSM, Lit. [7.29]

<u>Fig. 7.11a</u>. CsF: ω(q̲) [Ref. 7.30, Fig. 1], T = 80K, M: 11P-SM (full lines), 5P-FCM (dashed lines), Lit. [7.31-33]

<u>Fig. 7.11b</u>. CsF: D(ω) [Ref. 7.30, Fig. 3], M: 11P-SM, Lit. [7.31]

8. Cesium Chloride Structure Crystals

Crystals with this structure exhibit generally a complex dielectric behavior caused by the high polarizability of the constituent ions. Under pressure, several of them undergo phase transitions into the rock salt structure. Strong anharmonic effects are connected with these properties. The ammonium-halides are closely related to those with rock salt structures described in Chap.7.

Crystal	Figures showing	
	Dispersion curve $\omega(\underline{q})$	Density of states $D(\omega)$
CsCl	8.1a	8.1b
CsBr	8.2a	8.2b
CsI	8.3a	8.3b
NH_4Cl	8.4a	8.4b
ND_4Cl	8.5a	8.5b
ND_4Br	8.6	
TlCl	8.7	
TlBr	8.8a	8.8b

Fig. 8.1a. CsCl: ω($\underline{q}$)
[Ref. 8.1, Fig. 1],
T = 78 K, M: 11P-SM,
Lit. [8.2-4,22]

Fig. 8.1b. CsCl: D(ω)
[Ref. 8.1, Fig. 2],
M: 11P-SM, T = 78 K
(full lines), T = 298 K
(dashed lines)

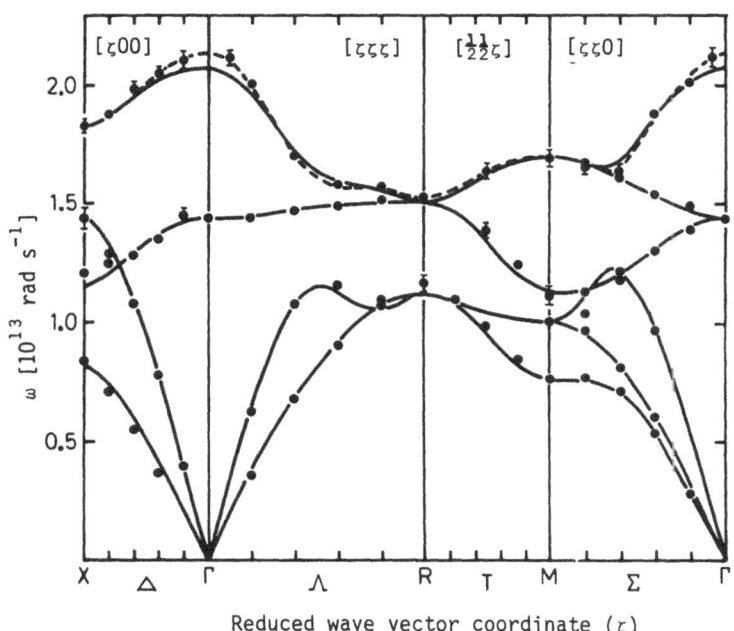

Reduced wave vector coordinate (ζ)

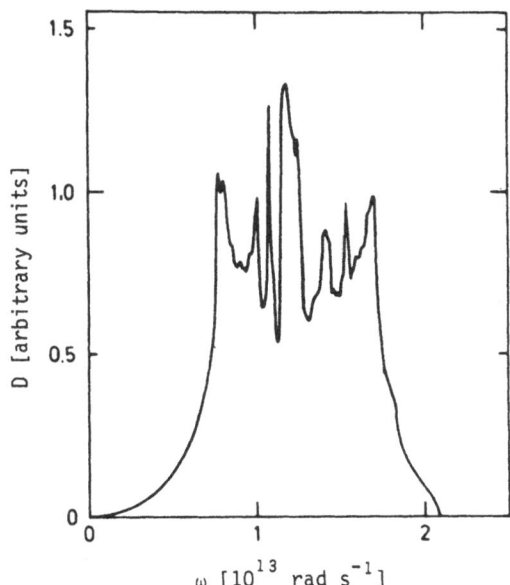

Fig. 8.2a. CsBr: ω(q) [Ref. 8.5, Fig. 1], T = 80 K, M: 11P-SM (full lines), 11P-BSM (dashed lines), Lit. [8.2,3,6-9,22]

Fig. 8.2b. CsBr: D(ω) [Ref. 8.5, Fig. 2], T = 80 K, M: 11P-SM, Lit. [8.9]

CsI

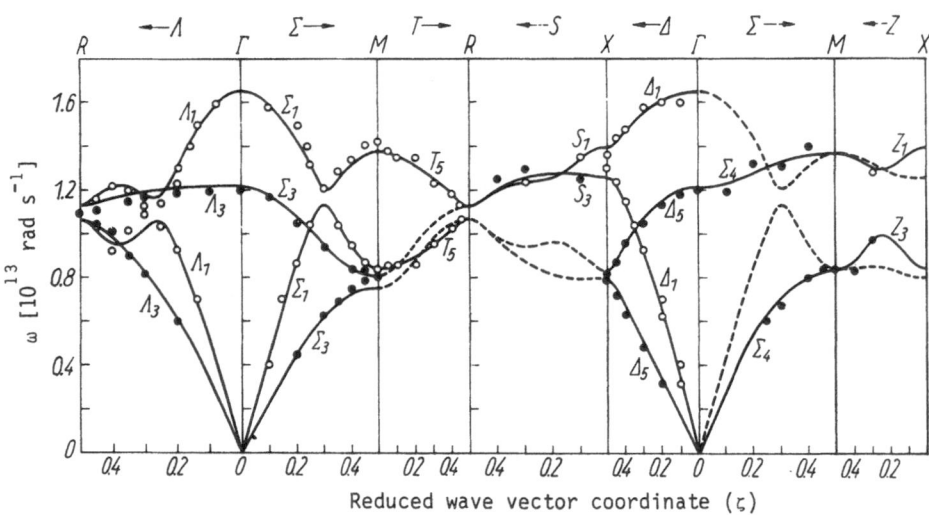

Fig. 8.3a. CsI: ω(q) [Ref. 8.10, Fig. 2], T = 293 K, M: 14P-SM,
Lit. [8.2,3,8,11,12,22]

Fig. 8.3b. CsI: D(ω)
[Ref. 8.10, Fig. 3],
T = 293 K, M: 14P-SM

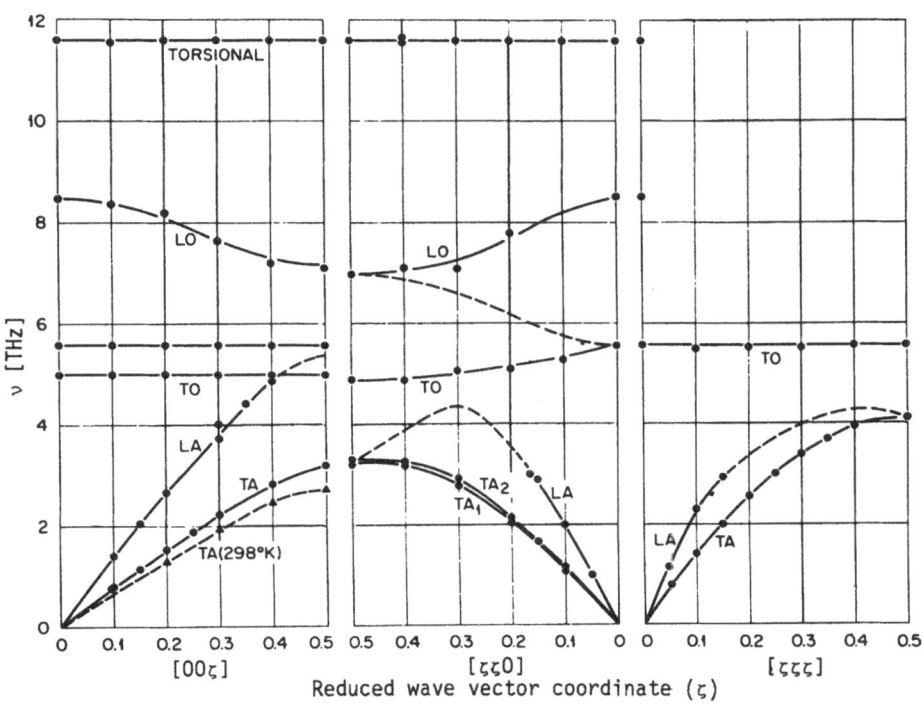

Fig. 8.4a. NH$_4$Cl: $\omega(\underline{q})$ [Ref. 8.13, Fig. 16], T = 78 K, M: -, Lit. [8.14-17]

Fig. 8.4b. NH$_4$Cl: D(ω)
[Ref. 8.14, Fig. 3],
T = 78 K, M: 5P-FCM,
Lit. [8.17]

ND₄Cl

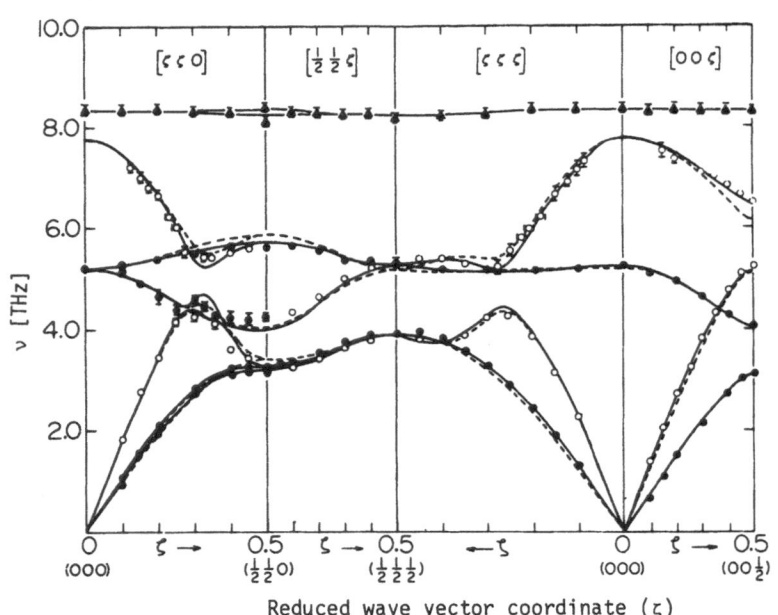

<u>Fig. 8.5a.</u> ND$_4$Cl: $\omega(\underline{q})$ [Ref.8.18, Fig.1], T = 85 K, M: 13P-SM (solid lines), 11P-FCM (dashed lines), Lit. [8.14,16,17,19-21]

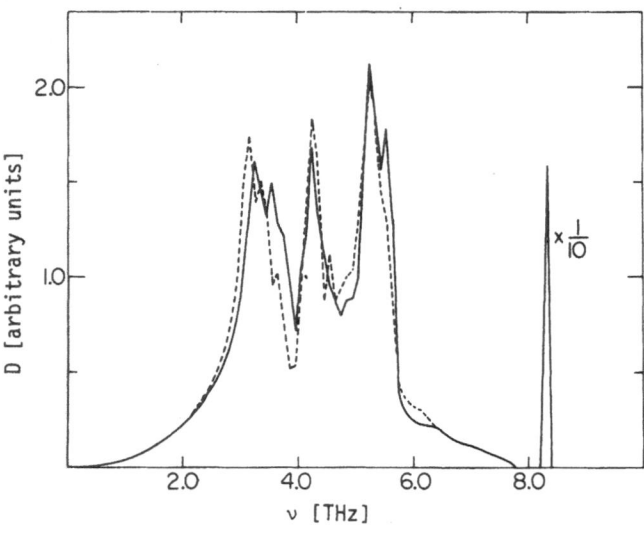

<u>Fig. 8.5b.</u> ND$_4$Cl: $D(\omega)$ [Ref. 8.18, Fig. 2], T = 85 K, M: 13P-SM (full lines), 16P-SM [18.16] (dashed lines), Lit. [8.16]

<u>Fig. 8.6.</u>. ND$_4$Br: $\omega(\underline{q})$ [Ref. 8.23, Fig. 2], T = 220 K, 295 K, M: RIM

TlCl

Fig. 8.7. TlCl: $\omega(\underline{q})$ [Ref. 8.24, Fig. 2]
T = 293 K, 80 K

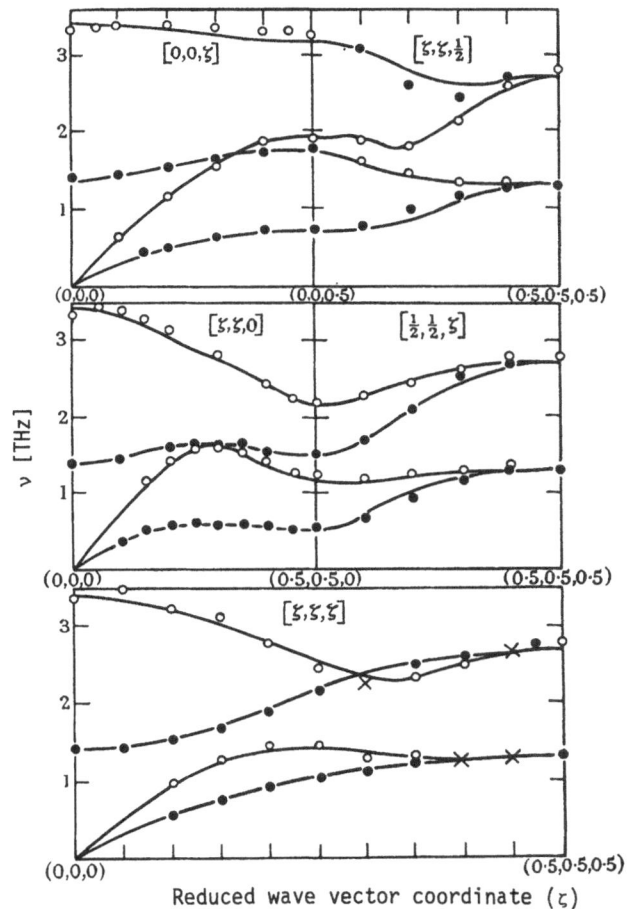

Fig. 8.8a. TlBr: ω(q)
[Ref. 8.25, Fig. 2],
T = 100 K, M: 14P-SM,
Lit. [8.26]

Reduced wave vector coordinate (ζ)

Fig. 8.8b. TlBr: D(ω)
[Ref. 8.25, Fig. 3a],
T = 100 K, M: 14P-SM

9. Diamond Structure Crystals

The covalency of the elemental semiconductors may be described in different ways in the different phonon models. In the shell model, COCHRAN [9.25] found that interionic core-shell interactions are important for the description of the flat TA branches in Si, Ge, and α-Sn. An alternative approach stems from the molecular picture using valence field forces [9.10]. The most pictorial way of description is the bond charge model (BCM) which in its general form [9.12] is a very powerful method to understand the mechanical properties of these crystals, although the dielectric properties are not well described.

The following systems are treated:

Crystal	Figures showing	
	Dispersion curve $\omega(\underline{q})$	Density of states $D(\omega)$
Diamond	9.1a	9.1b
Si	9.2a	9.2b
Ge	9.3a	9.3b
α-Sn	9.4a	9.4b

Diamond

Fig. 9.1a. Diamond: $\omega(\underline{q})$ [Ref. 9.1, Fig. 1], T = 296 K, M: 10P-SM, Lit. [9.2-12]

Fig. 9.1b. Diamond: $D(\omega)$ [Ref. 9.2, Fig. 3], T = 296 K, M: 10P-SM, Lit. [9.3,10,12]

Reduced wave vector coordinate

Fig. 9.2a. Si: ω(q) [Ref. 9.12, Fig. 3], T = RT, M: 4P-BCM (full lines),
6P-VFFM [9.10] (dashed lines) measurements [9.13,14], Lit. [9.3,7-19]

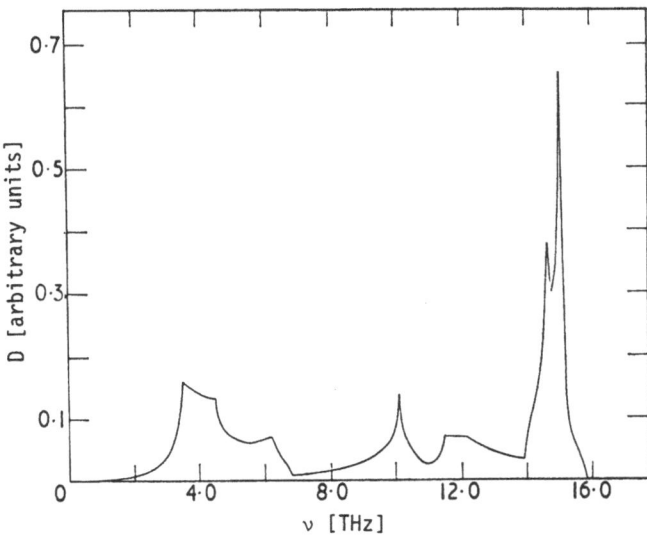

Fig. 9.2b. Si: D(ω) [Ref. 9.3, Fig. 2b], T = RT, M: 4P-BCM, Lit. [9.10,12]

Ge

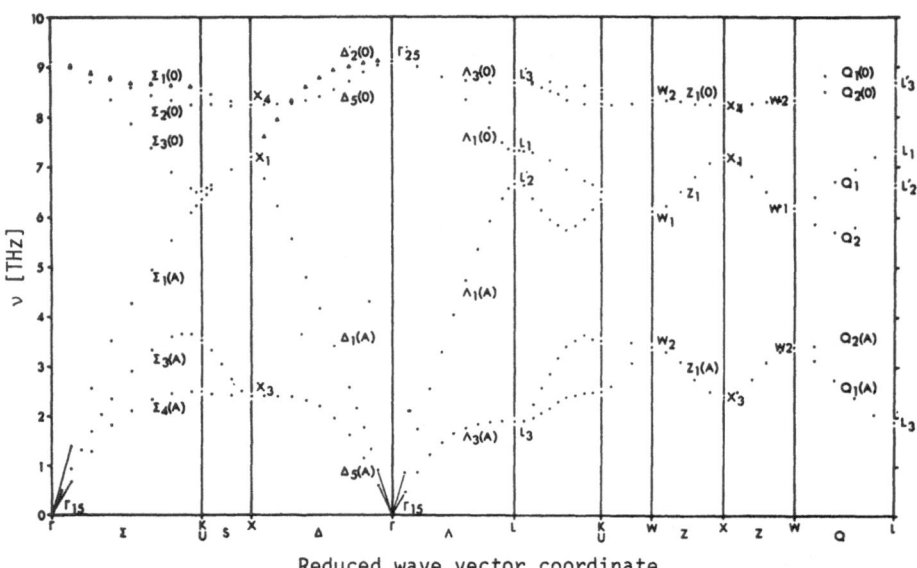

Fig. 9.3a. Ge: ω(q) [Ref. 9.20, Fig. 2], T = 80 K, Lit. [9.3,7-12,18,21-31]

Fig. 9.3b. Ge: D(ω) [Ref. 9.21, Fig. 3], T = 80 K, Lit. [9.3,10,12]

Fig. 9.4a. α-Sn: ω(q) [Ref. 9.32, Fig. 1], T = 90 K, M: 11P-SM,
Lit. [9.10,12,26,30,33-35]

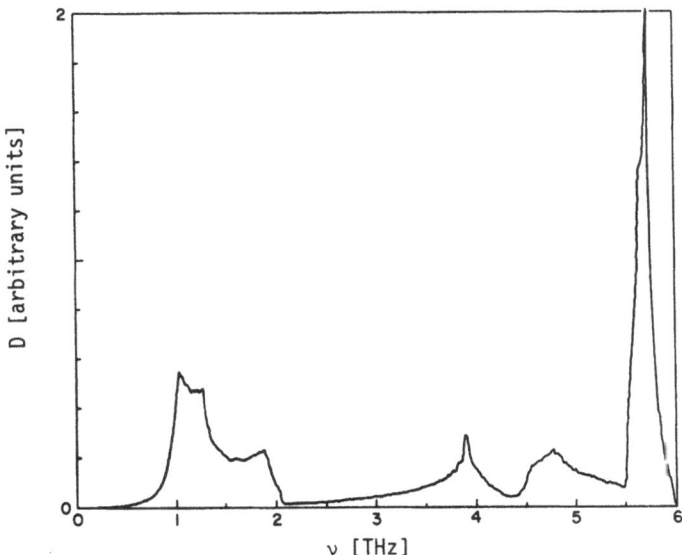

Fig. 9.4b. α-Sn: D(ω) [Ref. 9.33, Fig. 2a], T = 90 K, M: 11P-SM,
Lit. [9.10,12]

10. Zinc-Blende Structure Crystals

Crystals with this structure exhibit a complex mixture of covalent and ionic properties together with a generally high polarizability of the lattice ions. In crystals which are close to a phase transition to a more ionic structure (such as CuCl), strong anharmonic effects were observed [10.37,38]. The most successful model seems, at present, a combination of valence, ionic, and polarization forces including the overlap polarizability [10.11]. Another interesting is the asymmetric version of the bond charge model [10.12].

The following systems are treated:

Crystal	Figures showing		Crystal	Figures showing	
	Dispersion curve $\omega(\underline{q})$	Density of states $D(\omega)$		Dispersion curve $\omega(\underline{q})$	Density of states $D(\omega)$
SiC	10.1a	10.1b	ZnS	10.9a	10.9b
AlSb	10.2a	10.2b	ZnSe	10.10a	10.10b
GaP	10.3a	10.3b	ZnTe	10.11a	10.11b
GaAs	10.4a	10.4b	CdTe	10.12a	10.12b
GaSb	10.5a	10.5b	CuCl	10.13a	10.13b
InP	10.6a	10.6b	CuBr	10.14a	10.14b
InAs	10.7		CuI	10.15a	10.15b
InSb	10.8a	10.8b			

Fig. 10.1a. SiC: ω(q)
[Ref. 10.1, Fig. 2h]
M: 15P-DBM (full lines),
13P-DBM (dashed lines),
9P-DBM (dotted lines),
experimental data [10.2],
Lit. [10.3-6]

‾Reduced wave vector coordinate (ζ)

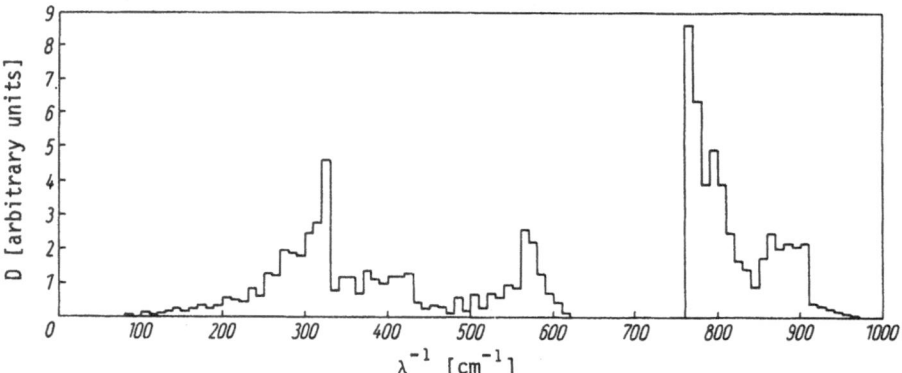

Fig. 10.1b. SiC: D(ω) [Ref. 10.1, Fig. 3h], M: 15P-DBM, Lit.[10.3,4]

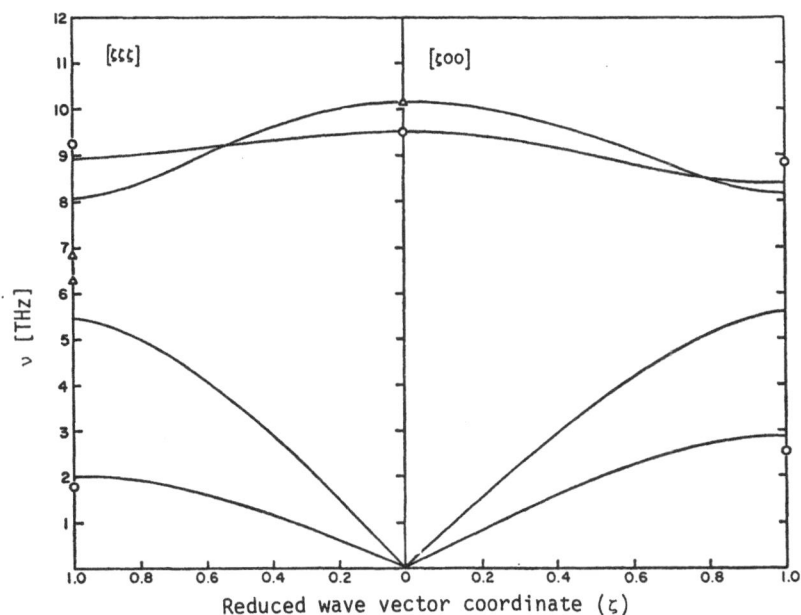

Fig. 10.2a. AlSb: ω(q) [Ref. 10.7, Fig. 2], M: 7P-RIM

Fig. 10.2b. AlSb: D(ω) [Ref. 10.7, Fig. 3], M: 7P-RIM

Fig. 10.3a. GaP: ω(q) [Ref. 10.8, Fig. 3], T = RT, M: 14P-SM, Lit. [10.9-12]

Fig. 10.3b. GaP: D(ω)
[Ref. 10.10, Fig. 3c],
M: 15P-DBM,
Lit. [10.9]

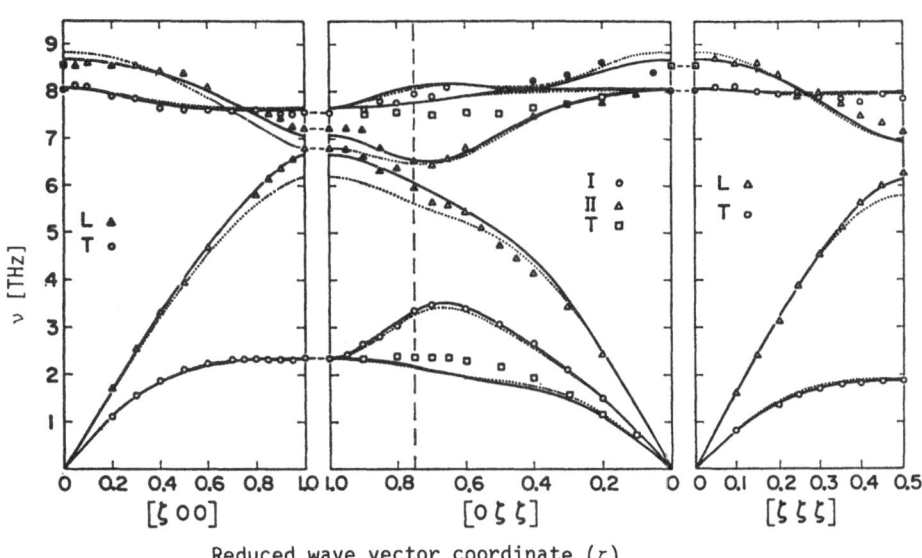

Fig. 10.4a. GaAs: ω(q) [Ref. 10.13, Fig. 1], T = 296 K, M: 14P-SM (full lines), 12P-SM (dashed lines), Lit. [10.7,10-13,15]

Fig. 10.4b. GaAs: D(ω) [Ref. 10.14, Fig. 2d], T = 296 K, M: 14P-SM, Lit. [10.7,10,13,14]

GaSb

Fig. 10.5a. GaSb: ω(q) [Ref. 10.16, Fig. 2], T = RT, M: 14P-SM,
Lit. [10.11,12,17]

Fig. 10.5b. GaSb: D(ω) [Ref. 10.16, Fig. 4], T = RT, M: 14P-SM

Fig. 10.6a. InP:
$\omega(\underline{q})$ [Ref. 10.18,
Fig. 4]
T = RT, M: -,
Lit. [10.19,20]

Fig. 10.6b. InP:
$D(\omega)$ [Ref. 10.19,
Fig. 7], T = RT,
M: 11P-RIM,
Lit. [10.20]

InAs

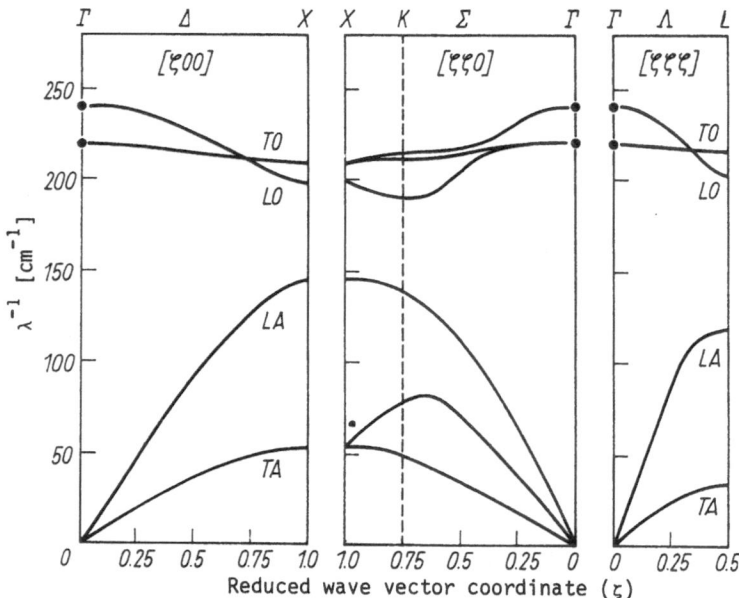

Fig. 10.7. InAs: $\omega_c(q)$ [Ref. 10.30, Fig. 6], M: 7P-RIM, Lit. [10.20]

Fig. 10.8a. InSb: ω(q) [Ref. 10.21, Fig. 2], T = RT, M: 14P-SM,
Lit. [10.10,12,20,22÷26]

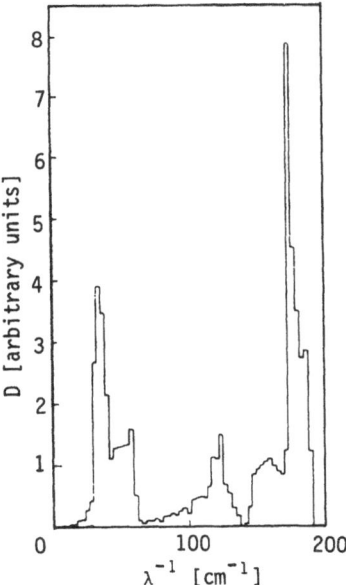

Fig. 10.8b. InSb: D(ω) [Ref. 10.10.,
Fig. 3e], M: 15P-DBM, Lit. [10.10,20,22]

ZnS

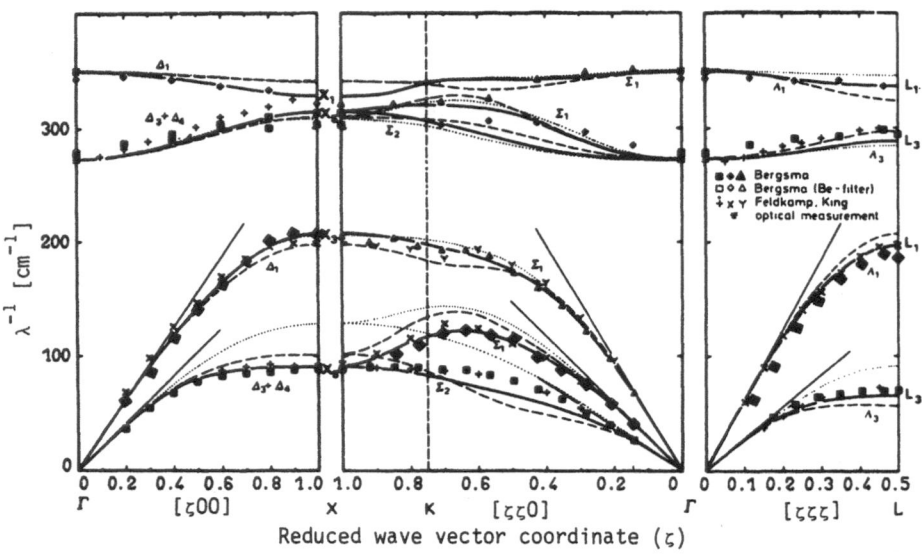

Fig. 10.9a. ZnS: ω(q) [Ref. 10.25, Fig. 1a], T = RT, M: 15P-DBM (full lines), 13P-DBM (dashed lines), 9P-DBM (dotted lines), measurements [10.27,28], Lit. [10.9,10,25,27-29,31,32]

Fig. 10.9b. ZnS: D(ω) [Ref. 10.10, Fig. 3a], M: 15P-DBM, Lit. [10.9,10,31,32]

Fig. 10.10a. ZnSe: ω(q) [Ref. 10.10, Fig. 2b], T = RT, M: 15P-DBM (full lines)
13P-DBM (dashed lines), 9P-DBM (dotted lines), measurements [10.33],
Lit. [10.11,23,33]

Fig. 10.10b. ZnSe: D(ω)
[Ref. 10.10, Fig. 3b],
M: 15P-DBM, Lit. [10.33]

ZnTe

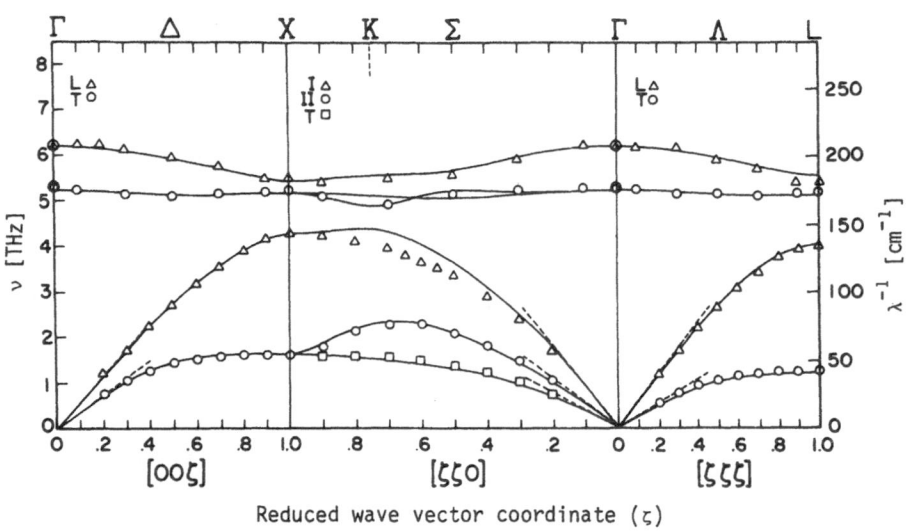

Fig. 10.11a. ZnTe: ω(q) [Ref. 10.32, Fig. 1], T = RT, M: 10P-VSM, Lit. [10.34]

Fig. 10.11b. ZnTe: D(ω) [Ref. 10.32, Fig. 2], T = RT, M: 10P-VSM, Lit. [10.34]

Fig. 10.12a. CdTe: ω(q) [Ref. 10.35, Fig. 1], T = 300 K, M: 14P-SM,
Lit. [10.34,36]

Fig. 10.12b. CdTe: D(ω) [Ref. 10.35, Fig. 3], M: 14P-SM, Lit. [10.34]

CuCl

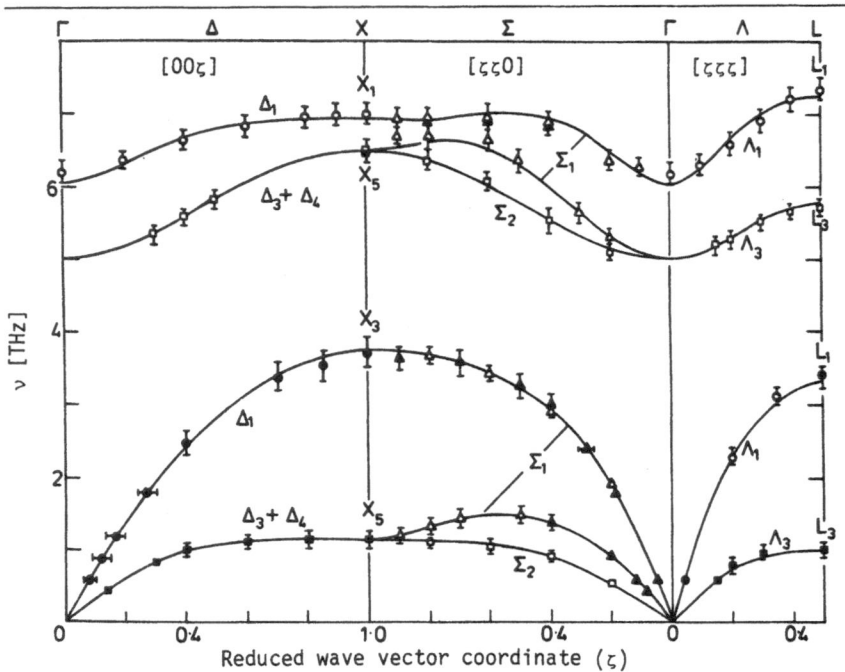

Fig. 10.13a. CuCl: ω(q) [Ref. 10.37, Fig. 1], T = 4.2 K, M: 14P-SM, Lit. [10.10,38,39]

Fig. 10.13b. CuCl: D(ω) [Ref. 10.37, Fig. 5], one phonon denisty (full lines), combined density of states (dashed lines), T = 4.2 K, M: 14P-SM, Lit. [10.10,39]

Fig. 10.14a. CuBr: ω(q) [Ref. 10.40, Fig. 3], T = 77 K, M: 14P-SM
(Parameter set I: dashed lines, parameter set II: full lines), Lit.[10.39,41]

Fig. 10.14b.
CuBr: [Ref.
10.40,Fig.4],
T = 77 K,
M: 14P-SM,
Lit. [10.39,
41]

Fig. 10.15a. CuI: ω(q)
[Ref. 10.42, Fig. 1],
T = 293 K, M: 7P-RIM,
Lit. [10.10,30,39]

Fig. 10.15b. CuI: D(ω)
[Ref. 10.42, Fig. 2],
T = 293 K, M: 7P-RIM,
Lit. [10.10,39]

11. Wurtzite Structure Crystals

The lattice dynamics of crystals with wurtzite structure belongs to the most difficult of diatomic crystals in general. As an example, the controversial situation in ZnO may be regarded. The delicate points are the 2NN forces which are different from those in crystals with zinc-blende structure (Chap. 10). Of particular interest is β-AgI for its relation to the fast ionic Ag conduction in α-AgI at high temperatures.

Crystal	Figures showing	
	Dispersion curve $\omega(\underline{q})$	Density of states $D(\omega)$
BeO	11.1a	11.1b
ZnO	11.2a 11.2b	
CdS	11.3a	11.3b
β-AgI	11.4a	11.4b

BeO

Fig. 11.1a. BeO: $\omega(\underline{q})$ [Ref. 11.1, Fig. 1], T = 300 K, Lit. [11.2-4]

Fig. 11.1b. BeO: $D(\omega)$ [Ref. 11.4, Fig.7], T = 293 K

Fig. 11.2a. ZnO: ω(q) [Ref. 11.5, Fig. 1], T = 298 K, M: 7P-SM, Lit. [11.7,8]

Fig. 11.2b. ZnO: ω(q) [Ref.11.6, Fig.1], T = 293 K, M: 10P-VSM, Lit.[11,7,8]

CdS

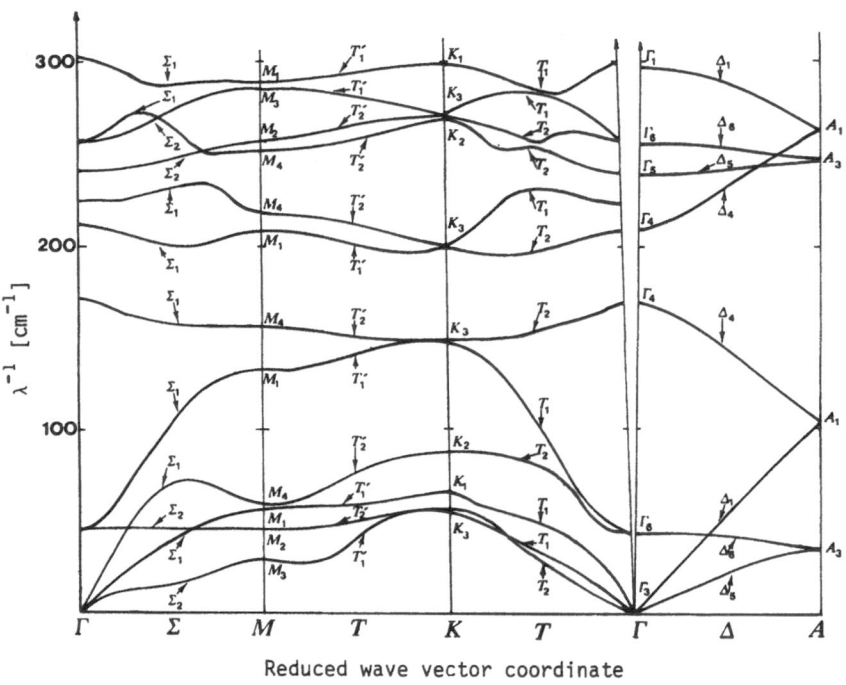

Fig. 11.3a. CdS: $\omega(\underline{q})$ [Ref. 11.9, Fig. 1], M: 10P-VDDM

Fig. 11.3b. CdS: $D(\omega)$ [Ref. 11.9, Fig. 2], M: 10P-VDDM

Fig. 11.4a. β-AgI: ω($\underline{q}$)
[Ref. 11.10, Fig. 3], T = 160 K,
M: 9P-VSM (full lines),
RIM (dotted lines)

Fig. 11.4b. β-AgI: D(ω)
[Ref. 11.10, Fig. 4], M: 9P-VSM

12. Fluorite Structure Crystals

CaF_2 and its structural homologues are, on the one side, normal ionic crystals, but on the other side have several members which show fast ionic conduction and, related to that, unusual anharmonic properties. The shell models used so far do not exhibit a clear relation to this latter feature in fluorite structured crystals.

The following systems are treated:

Crystal	Figures showing	
	Dispersion curve $\omega(\underline{q})$	Density of states $D(\omega)$
CaF_2	12.1a	12.1b
SrF_2	12.2a	12.2b
$SrCl_2$	12.3a	12.3b
BaF_2	12.4a	12.4b
PbF_2	12.5a	12.5b
Mg_2Sn	12.6	
Mg_2Pb	12.7a	12.7b
UO_2	12.8a	12.8b

Note added in proof: The dispersion curves of Na_2S (anti-fluorite structure) have been measured recently by W. Bührer and H. Bill and are reported in Helvetica Physica Acta *50*, 431 (1977).

CaF₂

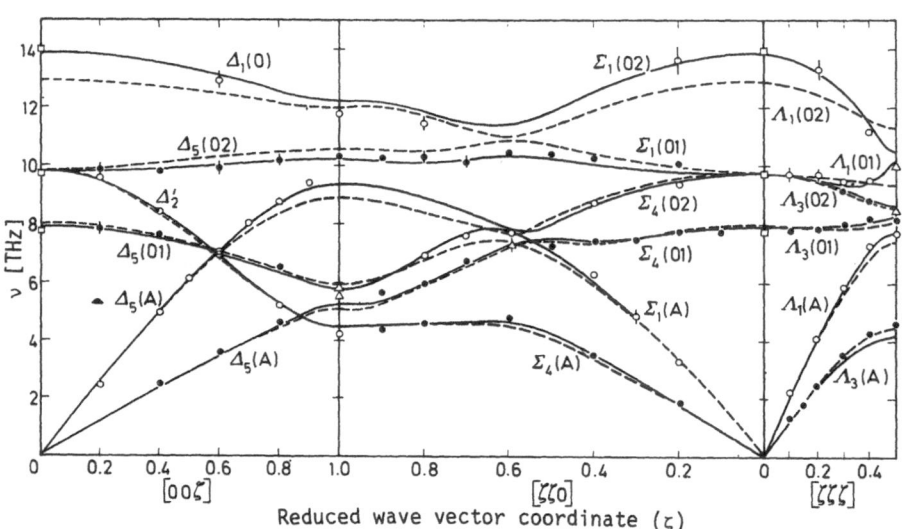

Fig. 12.1a. CaF$_2$: ω($\underline{q}$) [Ref. 12.1, Fig. 1], T = 295 K, M: 13P-SM (full lines), 7P-RIM (dashed lines)

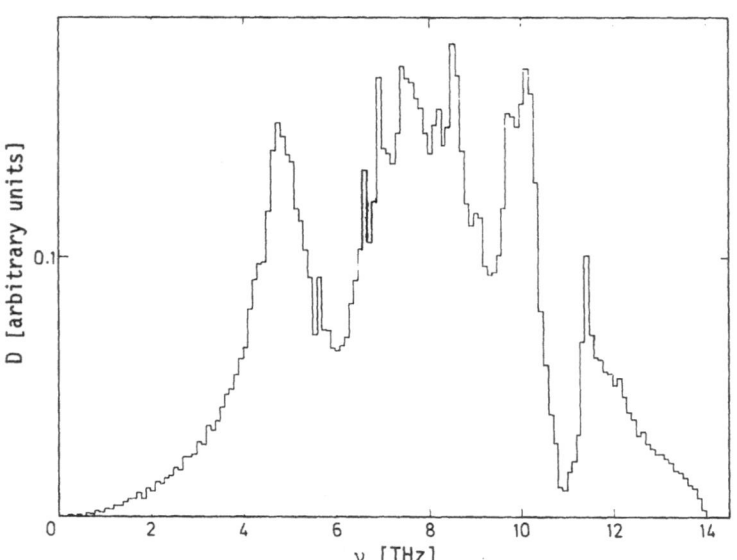

Fig. 12.1b. CaF$_2$: D(ω) [Ref. 12.1, Fig. 2], T = 295 K, M: 13P-SM

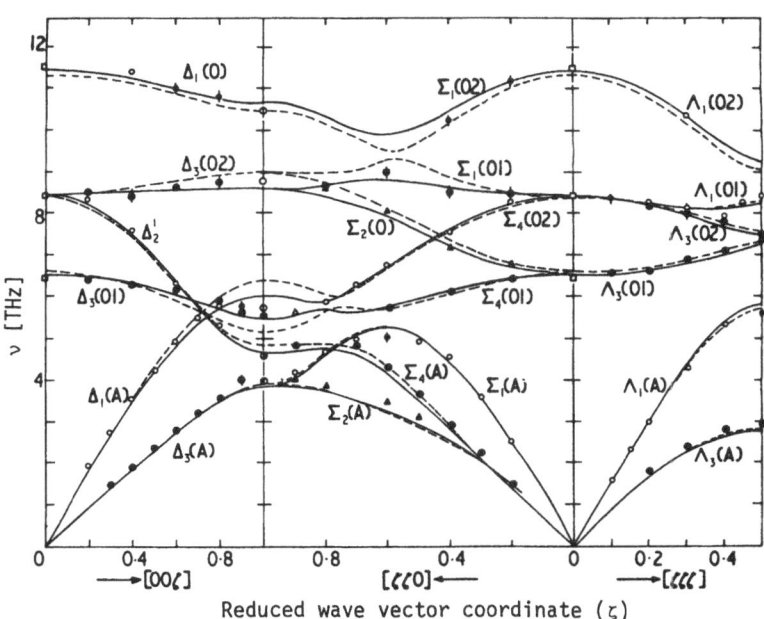

<u>Fig. 12.2a.</u> SrF$_2$: $\omega(\underline{q})$ [Ref. 12.2, Fig. 1], T = 295 K, M: 14P-SM (full lines), 10P-RIM (dashed lines)

<u>Fig. 12.2b.</u> SrF$_2$: $D(\omega)$ [Ref. 12.2, Fig. 2], T = 295 K, M: 14P-SM

Fig. 12.3a. SrCl$_2$: $\omega(\underline{q})$ [Ref. 12.3, Fig. 1], T = RT, M: 8P-SM

Fig. 12.3b. SrCl$_2$: $D(\omega)$ [Ref. 12.3, Fig. 2], T = RT, M: 8P-SM

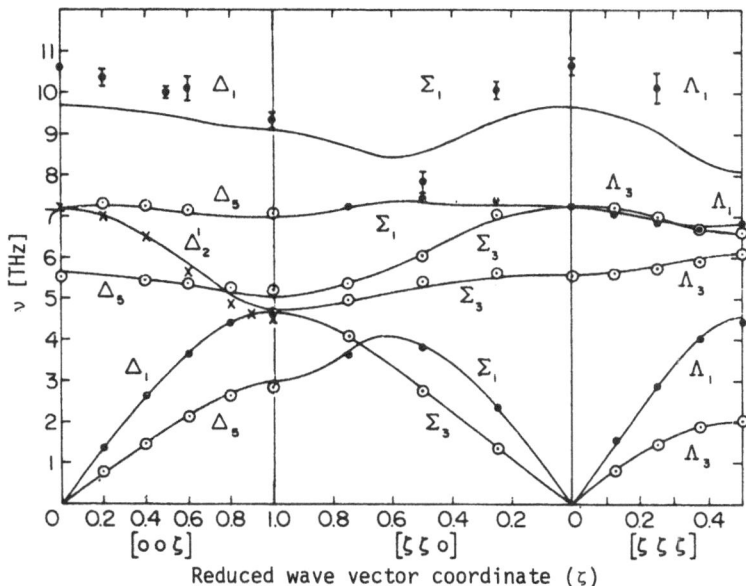

<u>Fig. 12.4a.</u> BaF$_2$: $\omega(\underline{q})$ [Ref. 12.4, Fig. 1], T = RT, M: 11P-SM

<u>Fig. 12.4b.</u> BaF$_2$: $D(\omega)$ [Ref. 12.4, Fig. 2], T = RT, M: 11P-SM

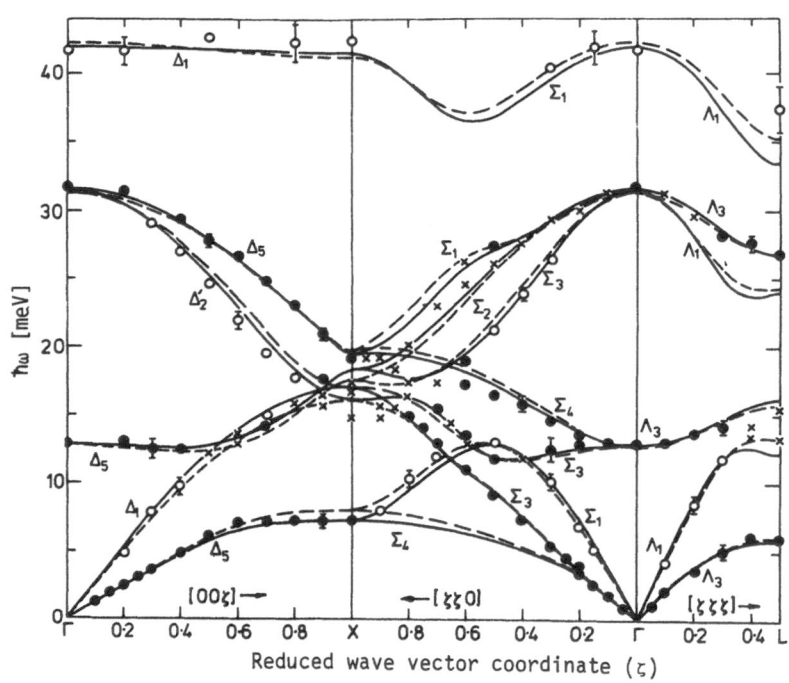

Fig. 12.5a. PbF$_2$: $\omega(\underline{q})$ [Ref. 12.5, Fig. 1], T = 10 K, M: 13P-SM (full lines),
8P-SM (dashed lines)

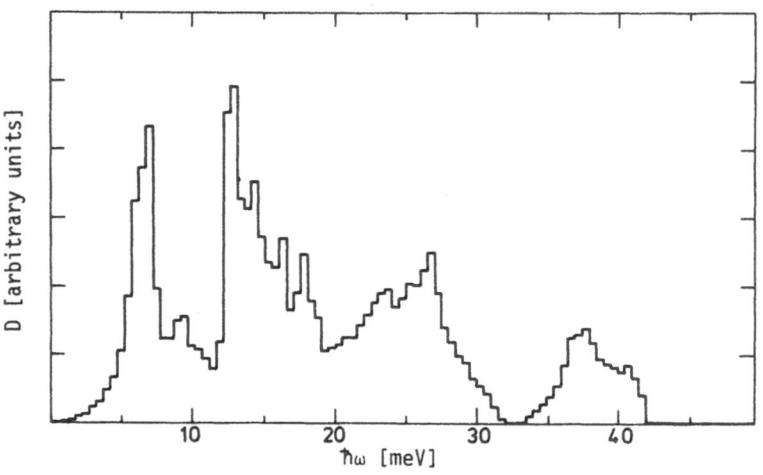

Fig. 12.5b. PbF$_2$: $D(\omega)$ [Ref. 12.5, Fig. 2], T = 10 K, M: 13P-SM

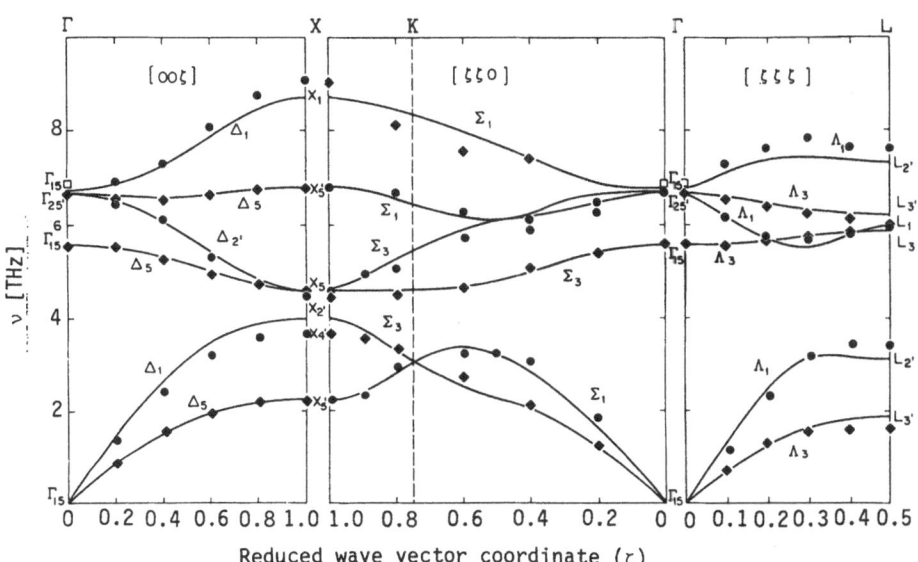

Fig. 12.6. Mg₂Sn: ω(q) [Ref. 12.6, Fig. 3], T = RT, M: 10P-SM

Mg₂Pb

Fig. 12.7a. Mg₂Pb:
ω(q) [Ref. 12.7,
Fig. 1], T = 293 K,
M: 9P-SM

Fig. 12.7b. Mg₂Pb:
D(ω) [Ref. 12.7,
Fig. 2], T = 293 K,
M: 9P-SM

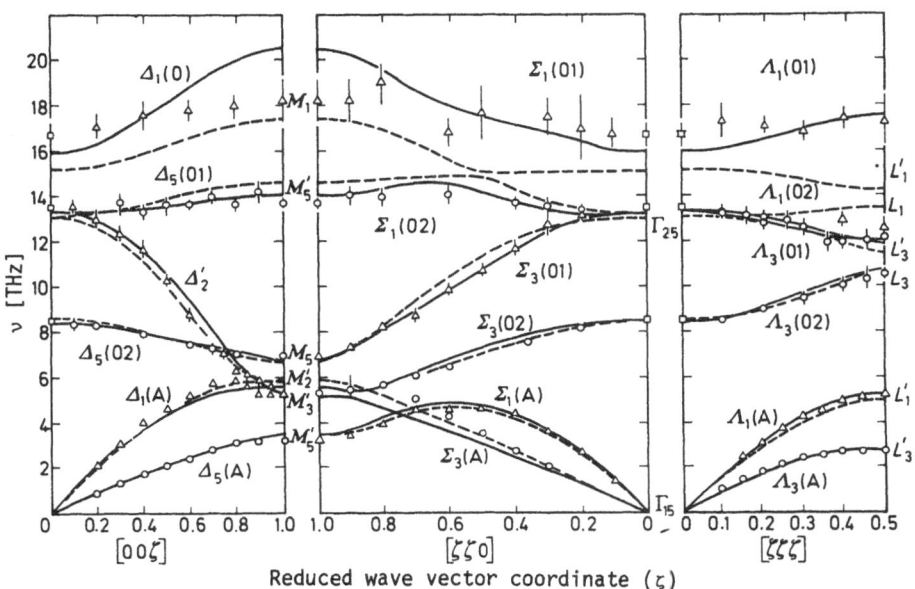

Fig. 12.8a. UO$_2$: ω(q) [Ref. 12.8, Fig. 4], T = 296 K, M: 11P-SM, (solid lines), 7P-RIM (dashed lines), Lit. [12.9]

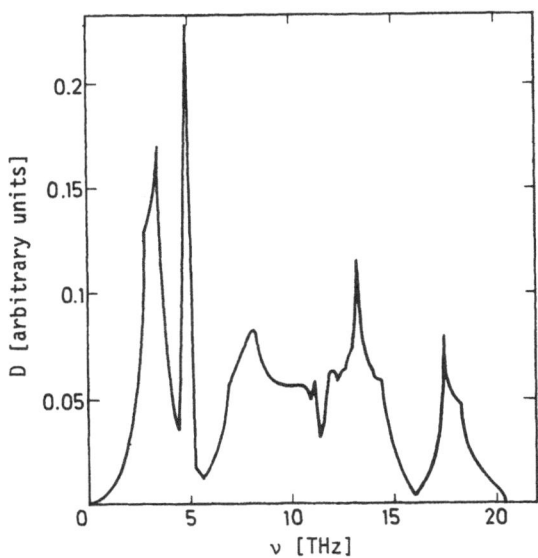

Fig. 12.8b. UO$_2$: D(ω) [Ref. 12.8, Fig. 5], M: 11P-SM, (solid lines), 7P-RIM (dashed lines)

13. Rutile Structure Crystals

The crystals in this group show many interesting features which one might expect in a relatively open structure, in particular phase transitions as a function of pressure and temperature. The description of the dispersion curves by shell models with many adjustable parameters is very unsatisfactory, at least in the oxides, while the fluorides exhibit a simpler behavior.

The following systems are treated:

Crystal	Figures showing	
	Dispersion curve $\omega(\underline{q})$	Density of states. $D(\omega)$
TiO_2	13.1a,b	13.1c
MgF_2	13.2a	13.2b
MnF_2	13.3a	13.3b
FeF_2	13.4	
CoF_2	13.5	
SnO_2	13.6	

TiO₂

Reduced wave vector coordinate (ζ)

<u>Fig. 13.1a.</u> TiO$_2$: $\omega(\underline{q})$ [Ref. 13.1, Fig. 5], T = RT, M: -, Lit. [13.2]

Fig. 13.1b. TiO$_2$: $\omega(\underline{q})$ [Ref. 13.1, Fig. 6], T = RT, M: -

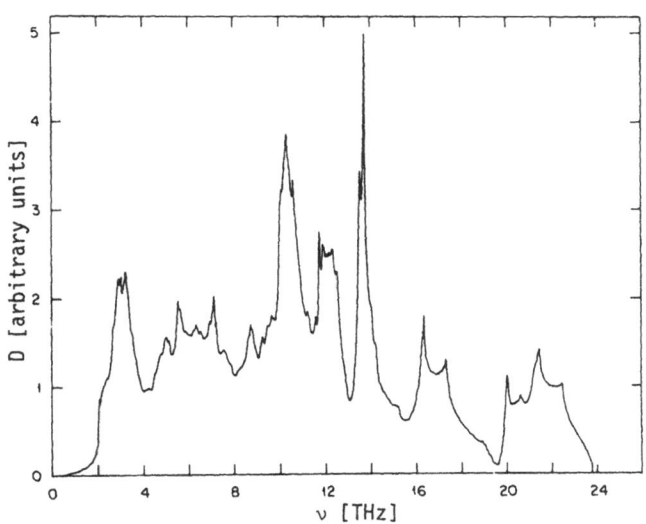

Fig. 13.1c. TiO$_2$: $D(\omega)$ [Ref. 13.1, Fig. 10], M: 18P-SM

MgF$_2$

Fig. 13.2a. MgF$_2$: $\omega(\underline{q})$
[Ref. 13.9, Fig. 2],
T = 296K, M: 14P-SM,
Lit. [13.3,5]

Reduced wave vector coordinate (ζ)

Fig. 13.2b. MgF$_2$:
D(ω) [Ref. 13.9,
Fig.3], M: 14P-SM

Fig. 13.3a. MnF$_2$: $\omega_c(\underline{q})$ [Ref. 13.4, Fig. 2], M: 11P-SM

Note added in proof: The dispersion curves measured by C.A. Rotter, J.G. Traylor, and H.G. Smith are presented in a review article by H.G. Smith and N. Wakabàyashi published in *Dynamics of Solids and Liquids by Neutron Scattering* edited by T. Springer and S. Lovesey; Topics in Current Physics, Vol. 3 (Springer, Berlin, Heidelberg, New York 1977) Fig. 2.8.

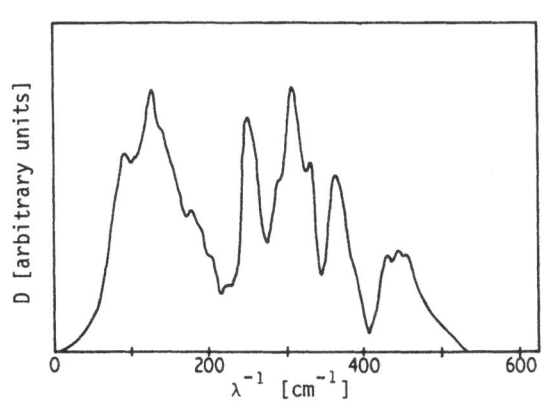

Fig. 13.3b. MnF$_2$: $D(\omega)$ [Ref. 13.4, Fig. 3a], M: 11P-SM

FeF$_2$

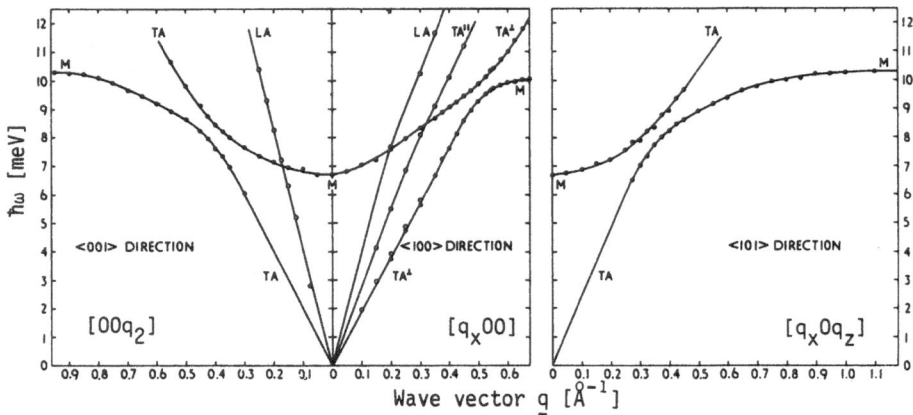

Fig. 13.4. FeF$_2$: $\omega(\underline{q})$ [Ref. 13.6, Fig. 1], T = 4.2K, M: –

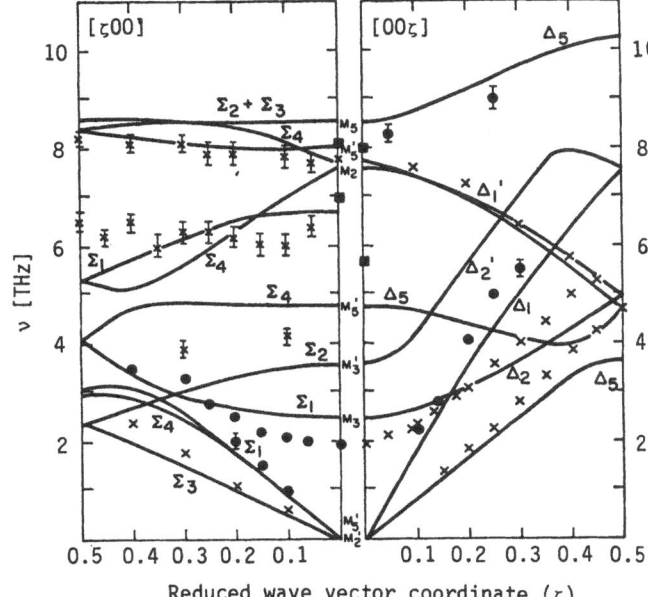

Fig. 13.5. CoF₂: ω(q) [Ref. 13.7, Fig. 5], T = 80K, M: 3P-RIM

SnO₂

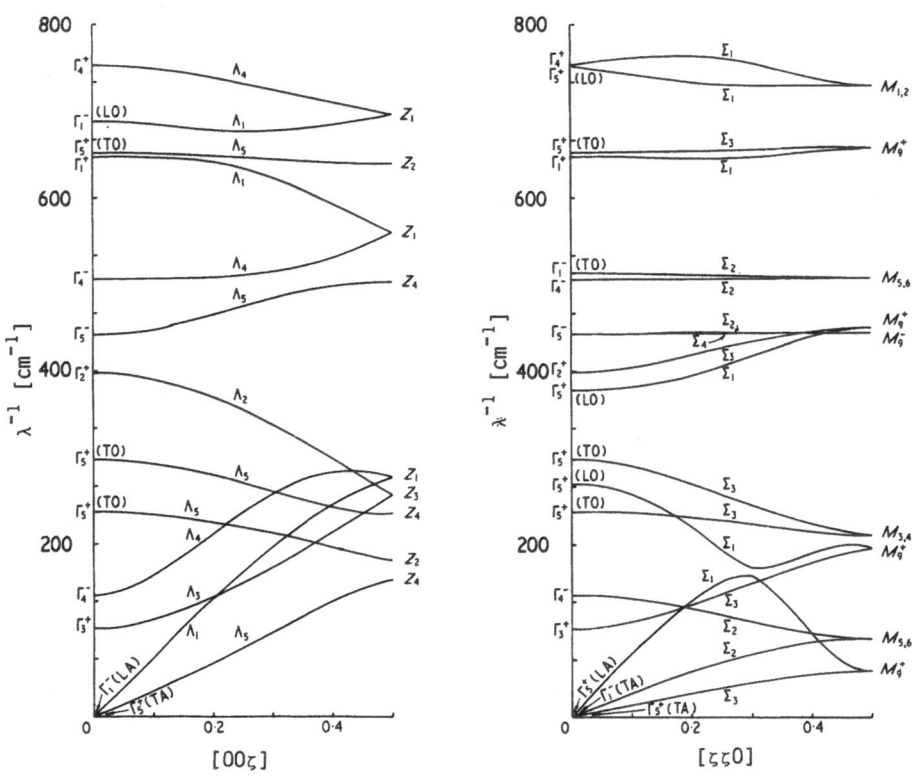

Reduced wave vector coordinate (ζ)

Fig. 13.6. SnO$_2$: $\omega_c(\underline{q})$ [Ref. 13.8, Fig. 4], M: 9P-RIM

14. ABO$_3$ and ABX$_3$ Crystals

In this chapter the phonon data of all measured perovskitic crystals are collected together with other crystals which belong to the chemical brutto formula ABX$_3$ (X = O,H,F). This, unavoidably, includes so-called molecular crystals like NaNO$_3$ but may be justified from a technical point of view. Since these crystals show many different types of phase transitions together with a complex dielectric and anharmonic behavior it is not surprising that the lattice dynamics of them is very complicated and offers many unsolved problems. An exception are the cubic perovskites where the pioneering work of COWLEY [14.12] and STIRLING [14.11] led to a basic understanding of interatomic forces in these substances. A recent microscopic model [14.7,8] emphasized the importance of the oxygen polarizability for the ferroelectric phase transition in ABO$_3$ perovskites.

The following systems are treated:

Crystal	Figures showing		Crystal	Figures showing	
	Dispersion curve $\omega(\underline{q})$	Density of states $D(\omega)$		Dispersion curve $\omega(\underline{q})$	Density of states $D(\omega)$
LiNbO$_3$	14.1		BaLiH$_3$		14.8
KNbO$_3$	14.2		LaAlO$_3$	14.9	
KTaO$_3$	14.3a-d		PbTiO$_3$	14.10a,b	
KMnF$_3$	14.4a,b		CsNiF$_3$	14.11	
SrTiO$_3$	14.5a	14.5b	KCoF$_3$	14.12	
SrLiH$_3$		14.6	CaCO$_3$	14.13	
BaTiO3 (cubic)	14.7a,b		NaNO$_3$	14.14	
			KNO$_3$	14.15	
BaTiO3 (tetrag.)	14.7c-e				

LiNbO₃

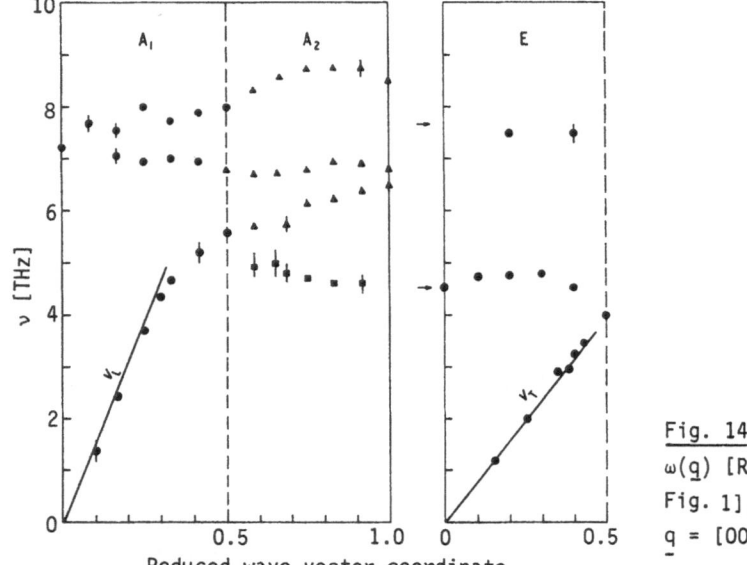

Fig. 14.1. LiNbO$_3$: $\omega(\underline{q})$ [Ref. 14.1, Fig. 1], T = RT, $\underline{q}$ = [00ζ]

Fig. 14.3a. KTaO₃:
ω(q̲) [Ref. 14.4,
Fig. 2],
T = 300K (open
circles), T = 20K
(solid circles),
M: -, Lit. [14.5-8]

Fig. 14.3b. KTaO₃: ω(q̲)
[Ref. 14.4, Fig. 3],
M: -, Lit. [14.5-8]

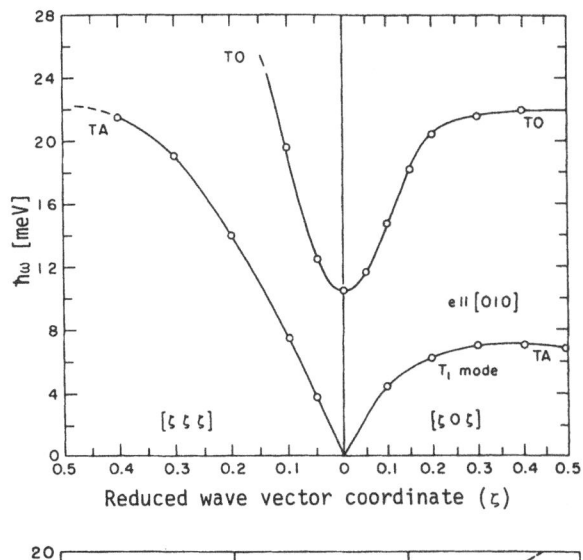

Fig. 14.3c. KTaO$_3$: ω($\underline{q}$) [Ref. 14.4, Fig. 4], T = 300 K, M: -, Lit. [14.5-8]

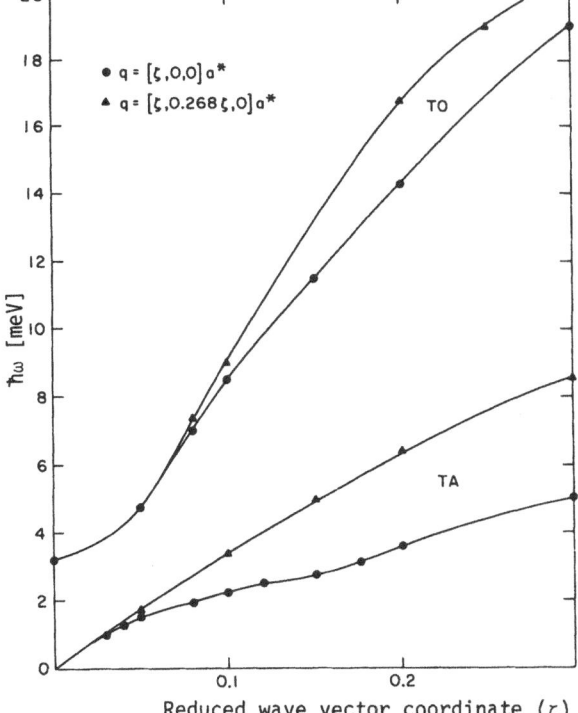

Fig. 14.3d. KTaO$_3$: ω($\underline{q}$) [Ref. 14.4, Fig. 5], T = 20 K, M: -, Lit. [14.5-8]

KMnF₃

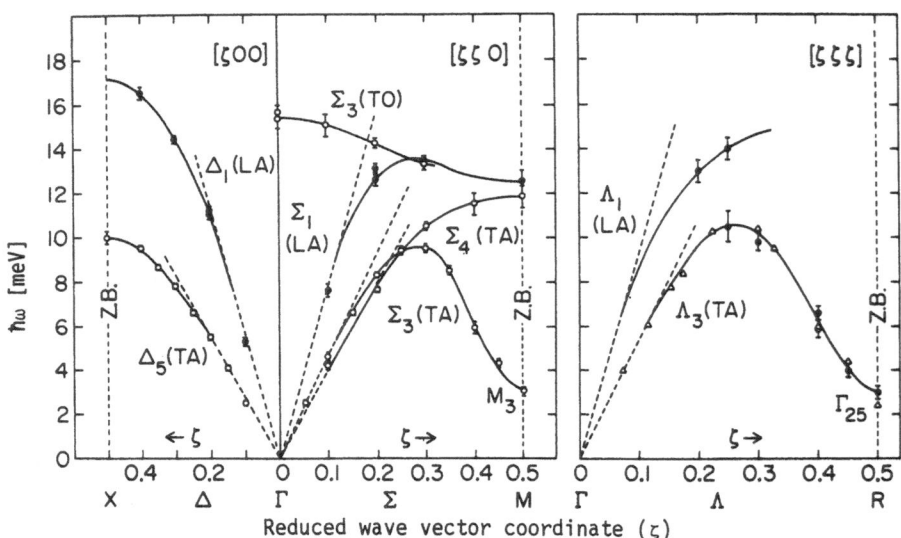

Fig. 14.4a. KMnF$_3$: $\omega(\underline{q})$ [Ref.14.9, Fig. 2], T = 295K, M: -, Lit. [14.6,10]

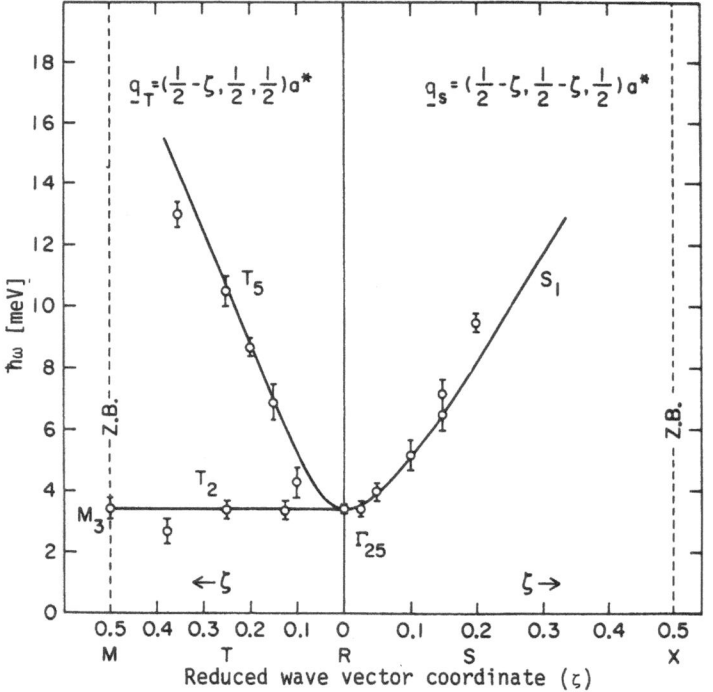

Fig. 14.4b.
KMnF$_3$: $\omega(\underline{q})$
[Ref. 14.9,
Fig. 3],
T = 295 K,
M: -, Lit.
[14.6,10]

Fig. 14.5a. SrTiO$_3$: ω(q̲)
[Ref. 14.11, Fig. 5],
T = 297K, M: 14P-SM,
crosses [14.12],
Lit. [14.6,13-18]

Fig. 14.5b. SrTiO$_3$: D(ω)
[Ref. 14.11, Fig. 6],
T = 297K (full lines),
90K (dashed lines),
M: 14P-SM

147

SrLiH₃

Fig. 14.6. SrLiH₃: D(ω) [Ref. 14.19, Fig. 4]

Fig. 14.7a. BaTiO₃
(cubic): ω(q)
[Ref. 14.20,
Fig. 1], q(ξ00),
T: 430°C (full
circles), 230°C
(open circles),
M: -,
Lit. [14.21-23]

Fig. 14.7b. BaTiO₃
(cubic): ω(q)
[Ref. 14.23,
Fig. 8],
T = 150°C, M: -
Lit. [14.20-22]

BaTiO₃ (tetragonal)

Fig. 14.7c. BaTiO$_3$
(tetragonal): ω($\underline{q}$)
[Ref. 14.24, Fig.6],
T = 22°C TA, TO,
e ‖ (100), M:-

Fig. 14.7d. BaTiO$_3$
(tetragonal): ω($\underline{q}$)
[Ref.14.24, Fig.8],
T = 22°C, TA, TO,
e ‖ (001), M: -

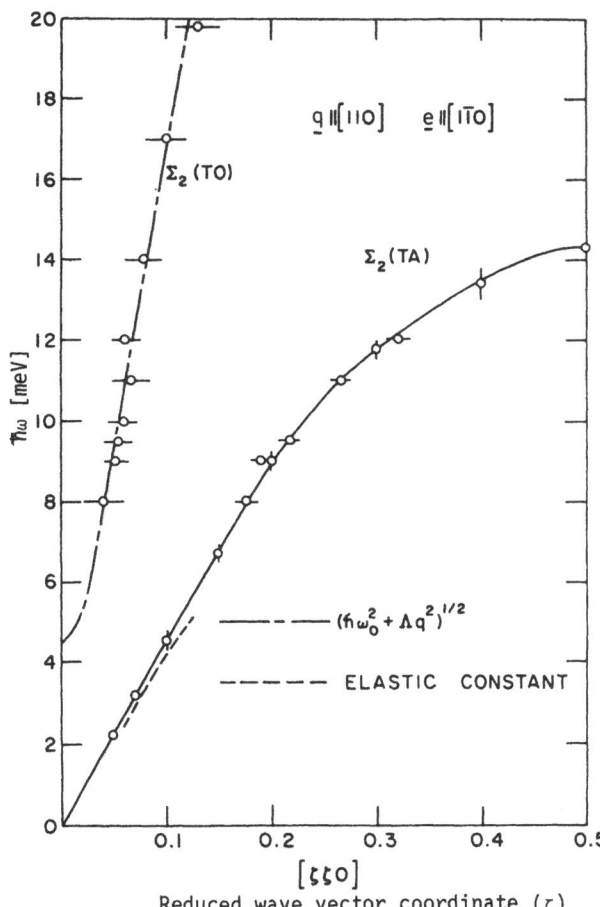

Fig. 14.7e. BaTiO$_3$
(tetragonal) $\omega(q)$
[Ref.14.24,Fig.3],
T = 22°C, TA, TO,
e || (110), M: -

BaLiH₃

Fig. 14.8. BaLiH$_3$: D(ω) [Ref. 14.19, Fig. 3]

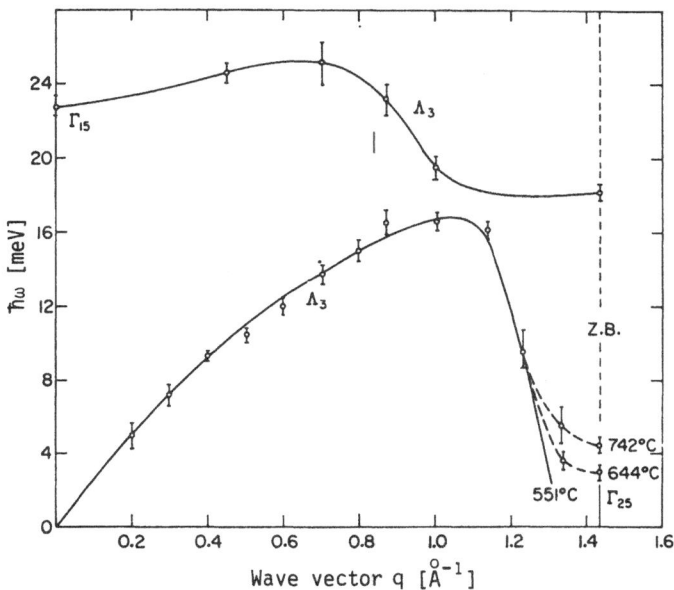

<u>Fig. 14.9.</u> LaAlO$_3$: ω(q̲) [Ref. 14.25, Fig. 3], T: 551°C, q̲ ‖ (111), M: -

PbTiO₃

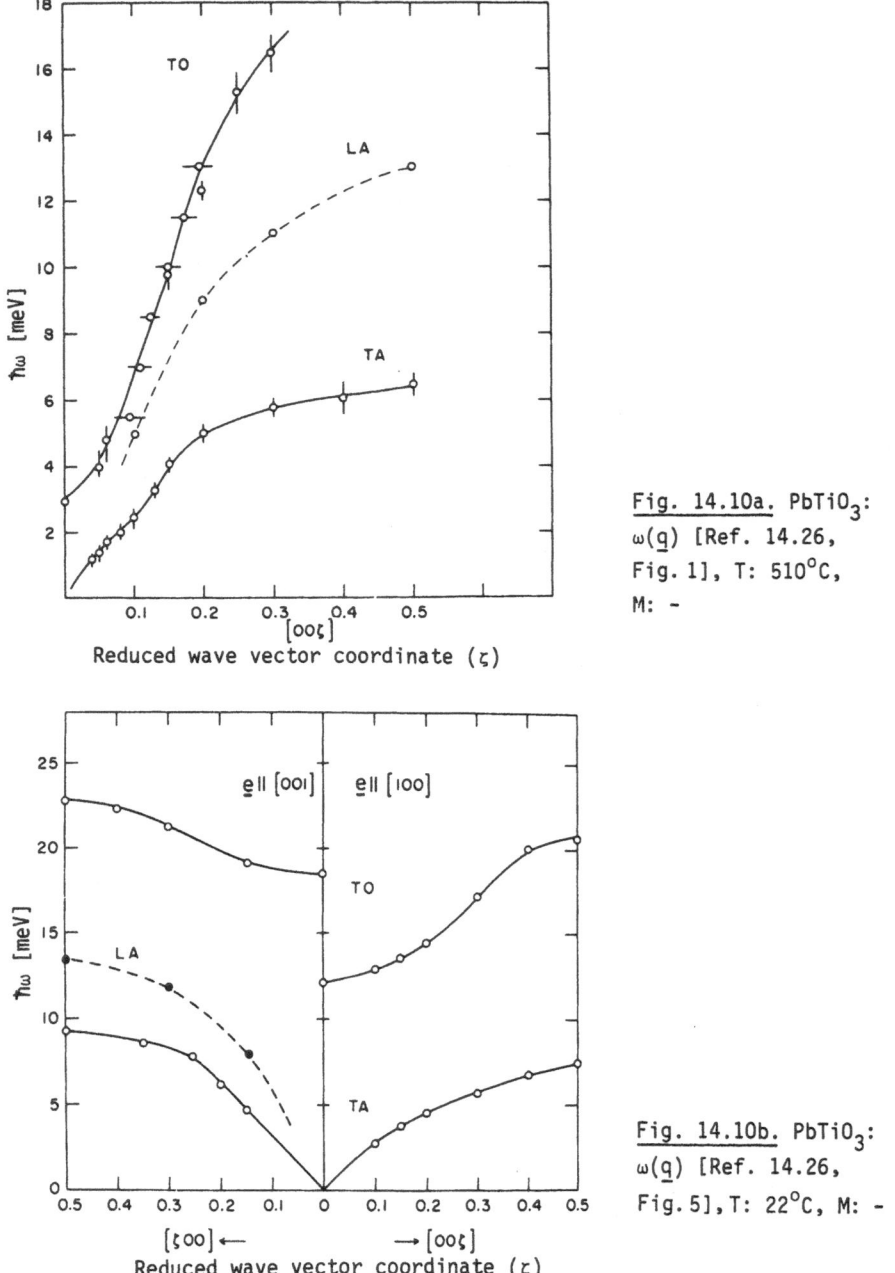

Fig. 14.10a. PbTiO₃:
ω(q) [Ref. 14.26,
Fig. 1], T: 510°C,
M: -

Fig. 14.10b. PbTiO₃:
ω(q) [Ref. 14.26,
Fig. 5], T: 22°C, M: -

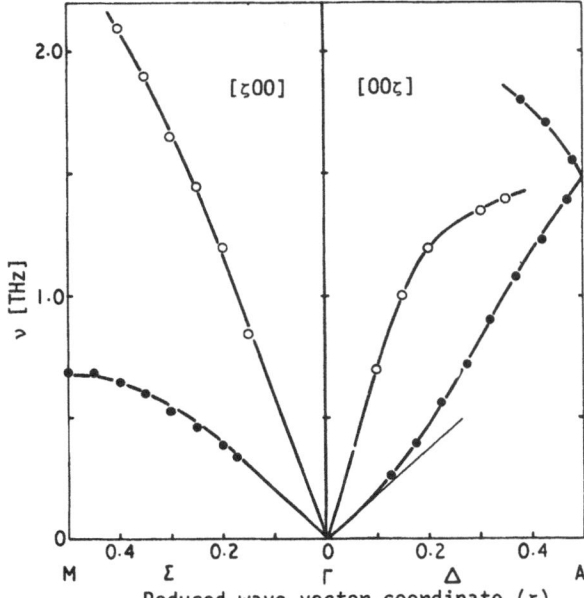

Fig. 14.11. CsNiF$_3$:
$\underline{\omega(q)}$ [Ref. 14.27,
Fig. 2], T = 85 K, M: -

KCoF₃

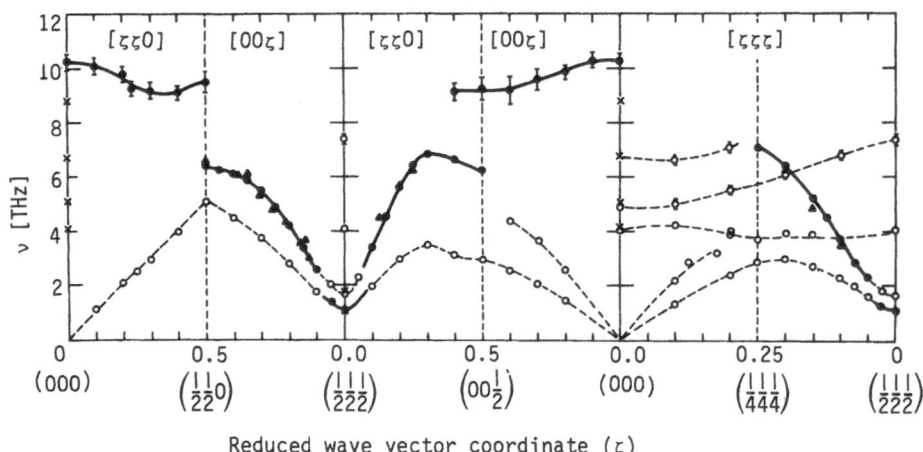

Fig. 14.12. KCoF$_3$: ω($\underline{q}$) [Ref. 14.28, Fig. 4], T = 22 K, M: -

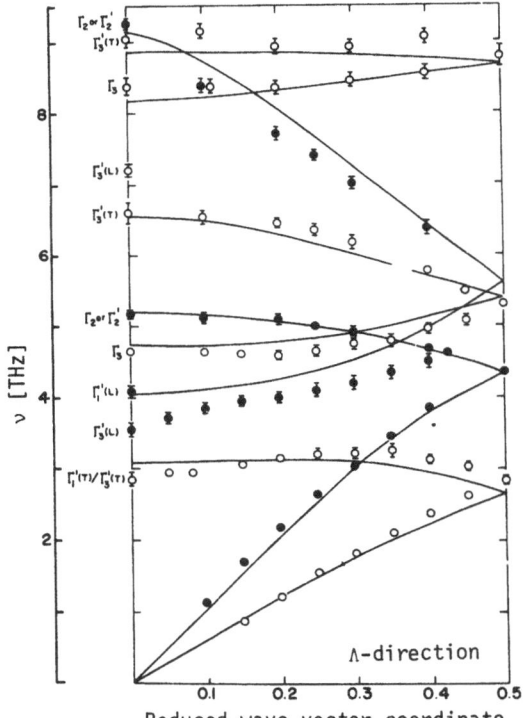

Fig. 14.13. CaCO₃: ω(q)
[Ref. 14.29, Fig. 1],
T = RT, M: 10P-SM

NaNO₃

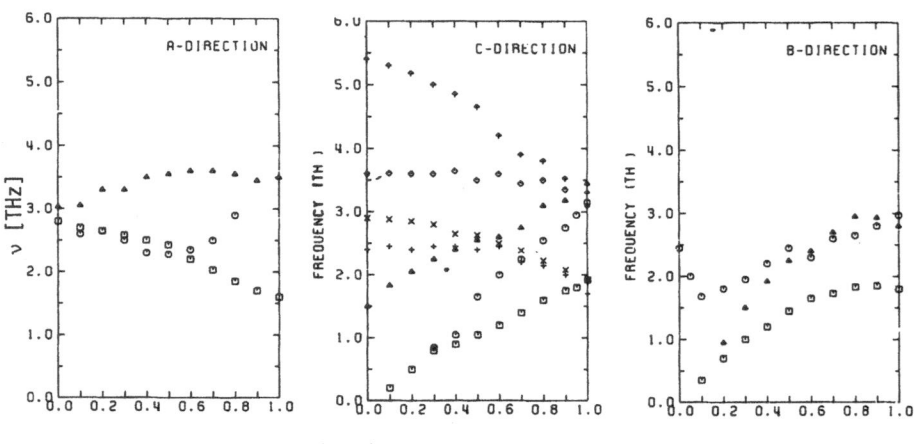

Reduced wave vector coordinate

Fig. 14.14. NaNO$_3$: $\omega(\underline{q})$ [Ref. 14.30, Fig. 1], T = RT, M: –

Fig. 14.15. KNO$_3$: ω($\underline{q}$) [Ref. 14.31, Fig. 1], T: RT, M: -, Lit. [14.32]

15. Layered Structure Crystals

Crystals with layer structures [15.31] are, at present, of particular interest. Their properties can very often be changed by intercalation, i.e., their structural anisotropy provides a short of chemical tool for a certain material "design" technique. The theoretical description is often done in terms of valence forces combined with axially symmetric forces. An important feature seems to be the structurally induced static dipoles [15.28,29] which lead to a quantitative understanding of the splitting between TO and LO branches. A full application of models used in cubic crystals should be very useful.

The following systems are treated:

Crystal	Figures showing		Crystal	Figures showing	
	Dispersion curve $\omega(\underline{q})$	Density of states $D(\omega)$		Dispersion curve $\omega(\underline{q})$	Density of states $D(\omega)$
Graphite	15.1a,b	15.1c	$CoCl_2$	15.8	
$NbSe_2$	15.2a,b		Bi_2Te_3	15.9a,b	
MoS_2	15.3a	15.3b	$SrGa_2$	15.10	
$SnSe_2$	15.4a,b		GaS	15.11a,b	15.11c
PbI_2	15.5		GaSe	15.12a	15.12b
$TiSe_2$	15.6a,b		GeS	15.13	
$FeCl_2$	15.7		TaS_2	15.14a,b	

Graphite

Fig. 15.1a. C (graphite): $\omega(q)$ [Ref. 15.1, Fig. 6], T = RT, M: 8P-ASM, Lit. [15.2-10]

Fig. 15.1b. C (graphite): $\omega(q)$ [Ref. 15.1, Fig. 8], M: 8P-ASM Lit. [15.2-10]

Fig. 15.1c. C (graphite): D(ω) [Ref.15.1, Fig.9], M: 8P-ASM, Lit.[15.2,4,5]

NbSe₂

Fig. 15.2a. NbSe₂: ω(q)
[Ref. 15.30, Fig. 12],
T = 300 K, M: -,
Lit. [15.11,12]

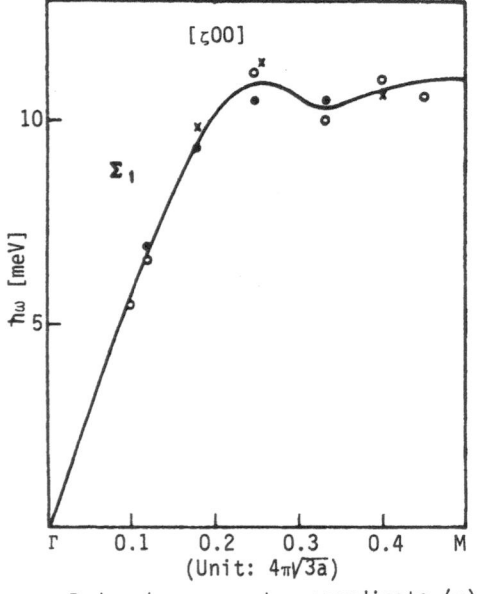

Fig. 15.2b. NbSe₂: ω(q)
Fig. 15.2b. NbSe₂: ω(q) [Ref.
15.30, Fig. 13], T = 300 K,
M: -, Lit. [15.11,12]

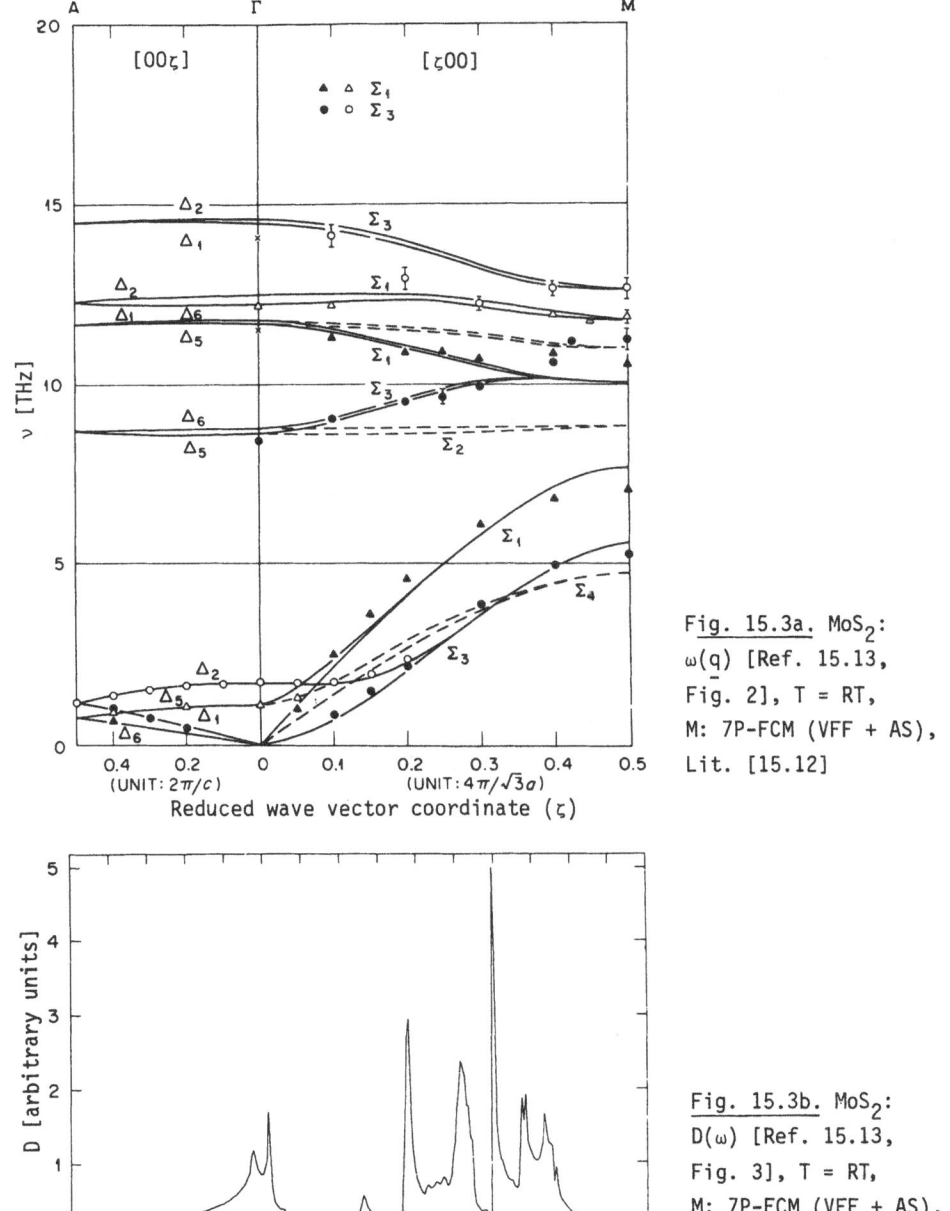

Fig. 15.3a. MoS$_2$:
$\omega(q)$ [Ref. 15.13,
Fig. 2], T = RT,
M: 7P-FCM (VFF + AS),
Lit. [15.12]

Fig. 15.3b. MoS$_2$:
D(ω) [Ref. 15.13,
Fig. 3], T = RT,
M: 7P-FCM (VFF + AS),
Lit. [15.12]

SnSe₂

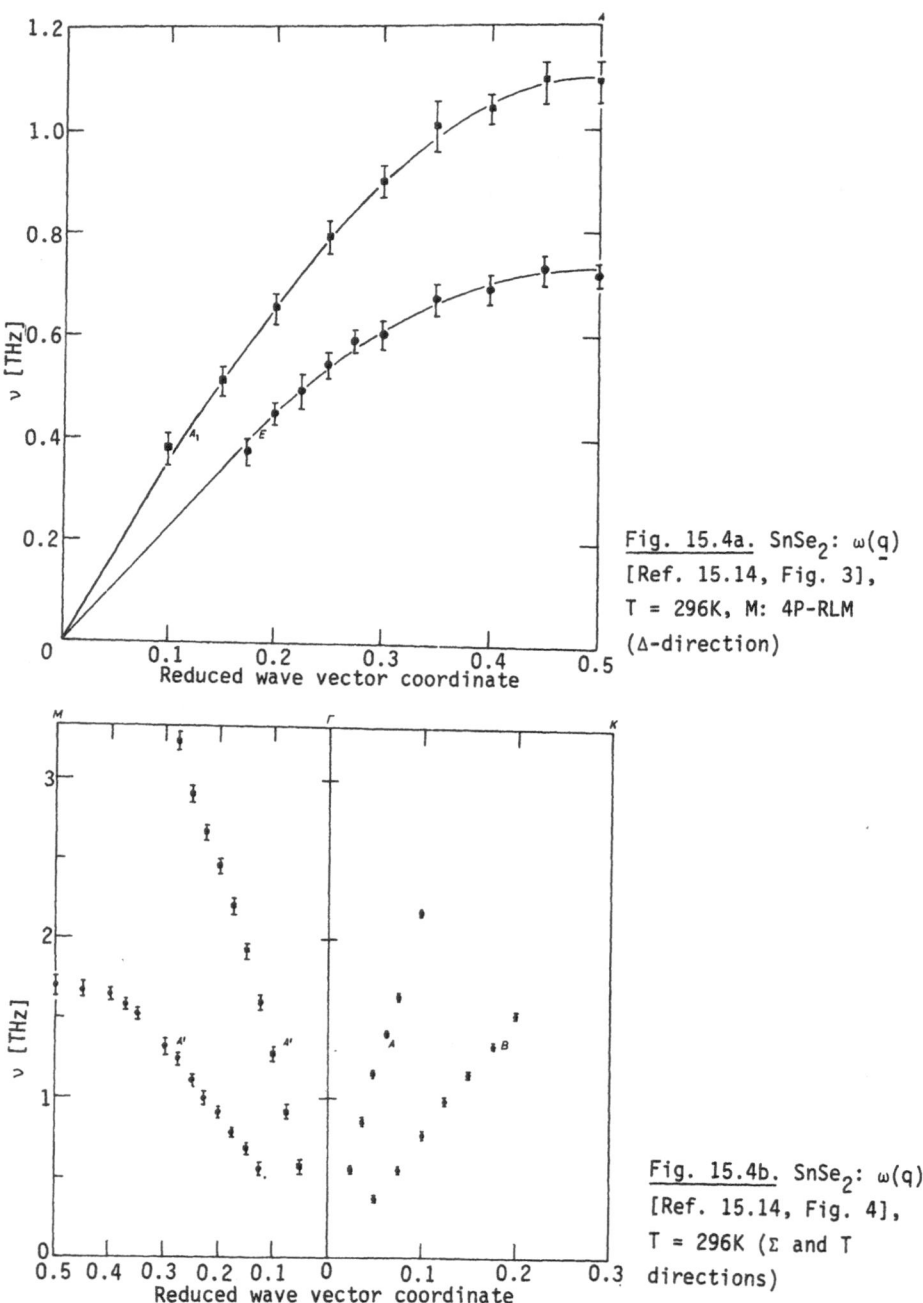

Fig. 15.4a. SnSe₂: ω(q)
[Ref. 15.14, Fig. 3],
T = 296K, M: 4P-RLM
(Δ-direction)

Fig. 15.4b. SnSe₂: ω(q)
[Ref. 15.14, Fig. 4],
T = 296K (Σ and T
directions)

Fig. 15.5. PbI$_2$: $\omega(\underline{q})$ [Ref. 15.15, Fig. 4], T = RT, M: -, Lit. [15.12,29]

TiSe$_2$

Fig. 15.6a. TiSe$_2$: $\omega(q)$ [Ref. 15.16, Fig. 2], T = RT, M: -

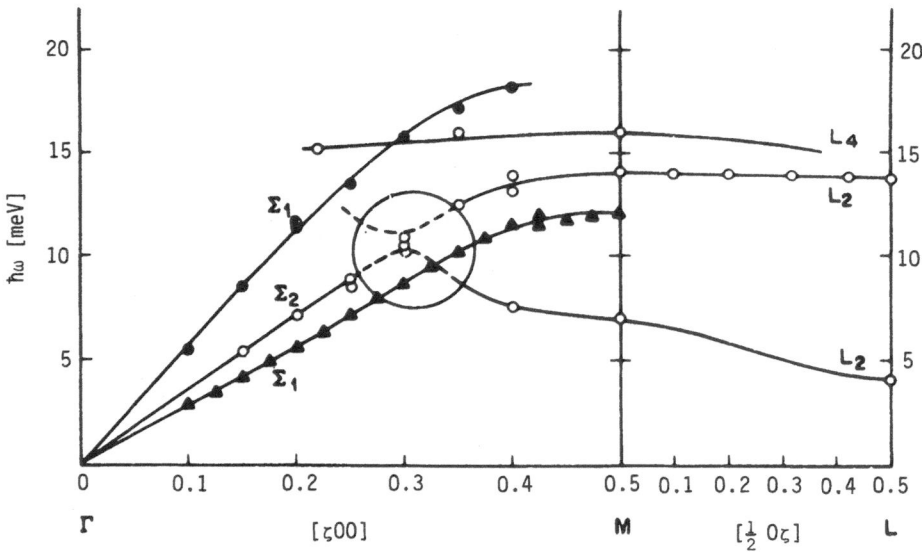

Fig. 15.6b. TiSe$_2$: $\omega(q)$ [Ref. 15.17, Fig. 4], T = RT, M: -

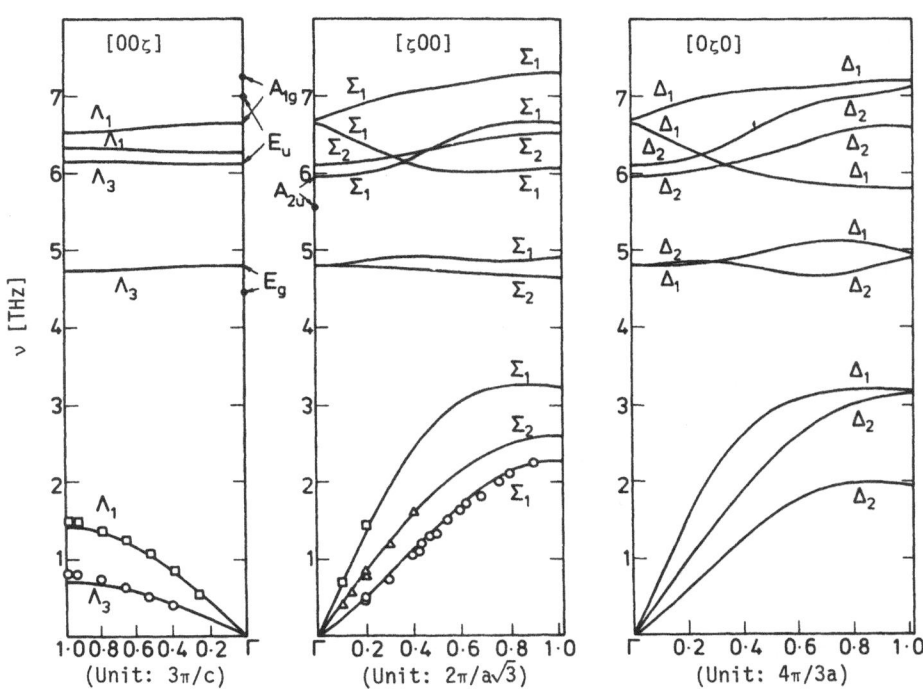

Fig. 15.7. FeCl$_2$: $\omega(q)$ [Ref. 15.18, Fig. 3b], measurements [15.19], T = RT,
M: 8P-SM (static displaced shells included), Lit. [15.20]

CoCl$_2$

Fig. 15.8. CoCl$_2$: $\omega(\underline{q})$ [Ref. 15.21, Fig. 1], T = RT, Lit. [15.20], M: –

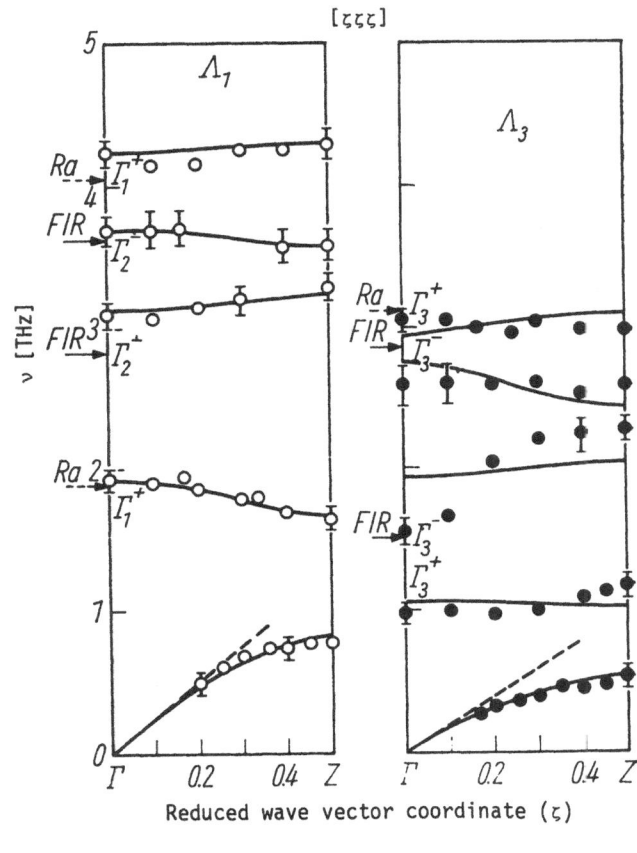

Fig. 15.9a. Bi$_2$Te$_3$:
$\omega(\underline{q})$ [Ref. 15.22,
Fig. 3], T = 77K,
M: 12P-RLM modes
propagating along
the trigonal axis

Fig. 15.9b. Bi$_2$Te$_3$:
$\omega(\underline{q})$ [Ref. 15.22,
Fig. 4], T = 77K,
modes propagating
along the bisectrix

SrGa₂

Fig. 15.10. SrGa₂: ω(q) [Ref. 15.23, Fig. 1], T = 20°C

Fig. 15.11a. GaS: ω(q) [Ref. 15.24, Fig. 2], T = 296 K, M: 9P-ASM, Lit. [15.25]

Fig. 15.11b. GaS: ω(q) [Ref. 15.24, Fig.4], M: 9P-ASM Lit. [15.25]

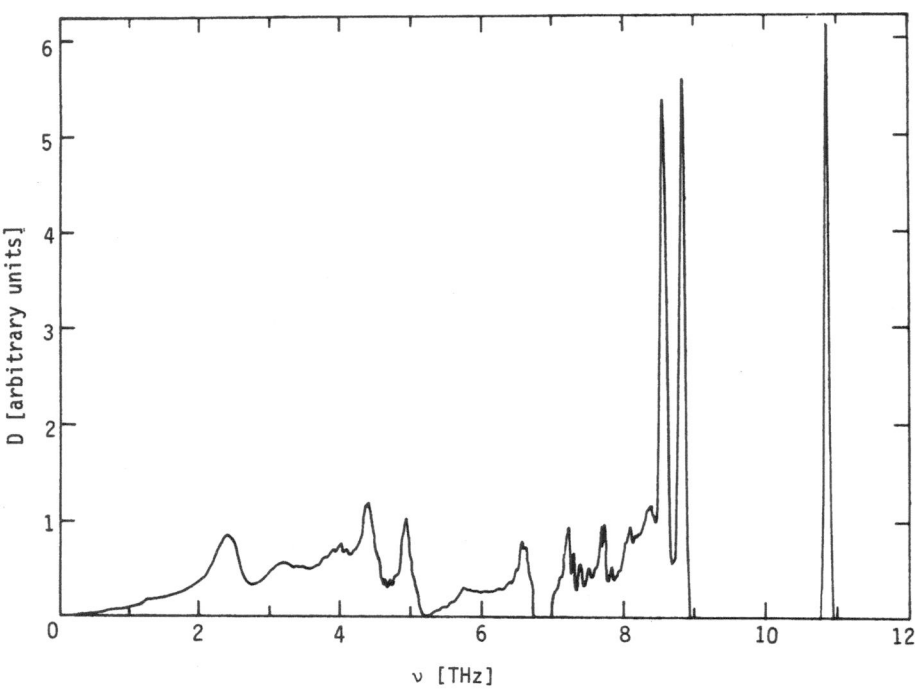

Fig. 15.11c. GaS: D(ω) [Ref. 15.24, Fig. 5], M: 9P-ASM

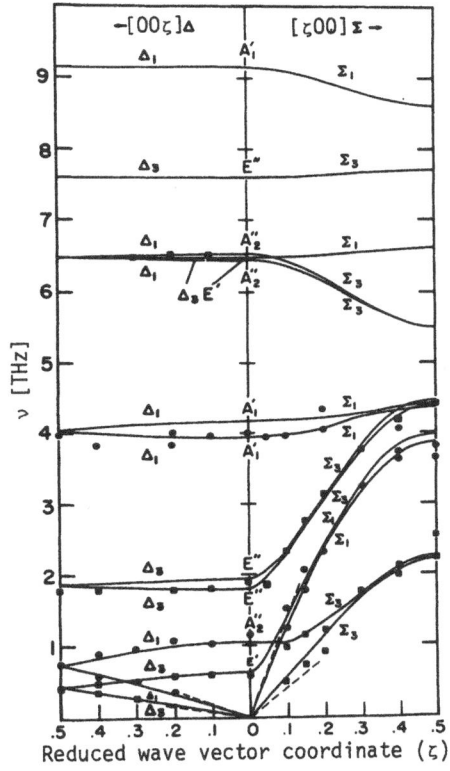

Fig. 15.12a. GaSe: ω(q) [Ref. 15.26, Fig. 3], T = 100 K, 9P-FCM

Fig. 15.12b. GaSe: D(ω) [Ref. 15.26, Fig. 5], T = 100 K, 9P-FCM

GeS

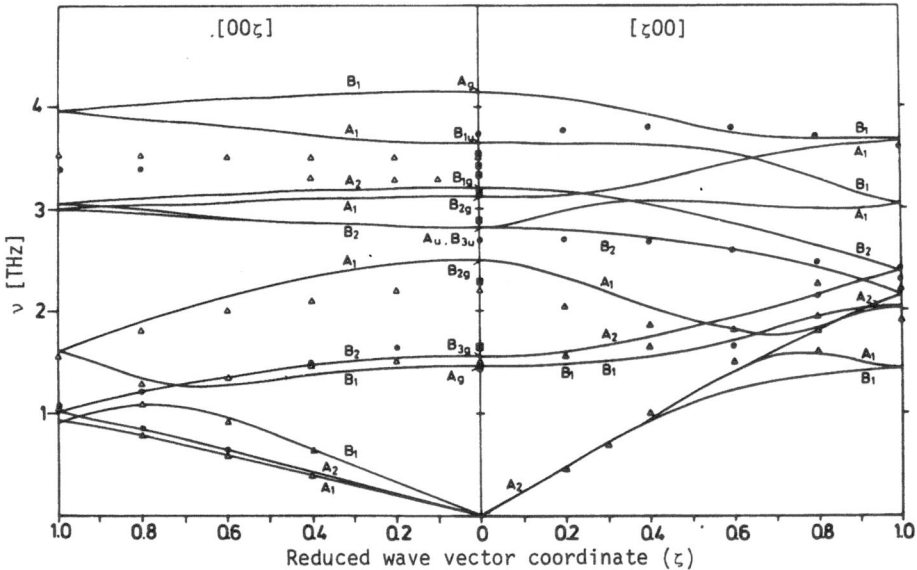

Fig. 15.13. GeS: ω(q) [Ref. 15.27, Fig. 1], T = RT, M: 4P-VFFM

Fig. 15.14a. TaSe$_2$: $\omega(\underline{q})$ [Ref. 15.30, Fig. 11], T = 300 K, M: -

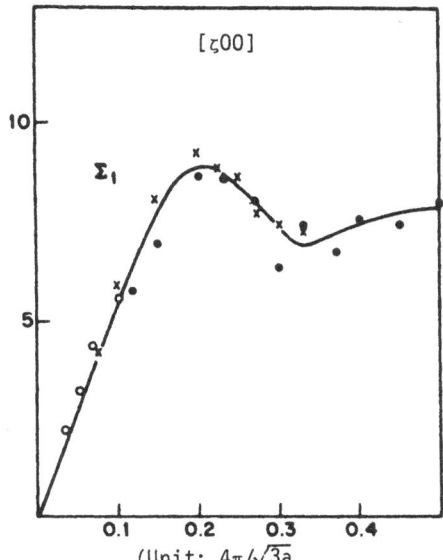

Fig. 15.14b. TaSe$_2$: $\omega(q)$ [Ref.15.30, Fig. 13], T = 300 K, M: -

16. Other Low-Symmetry Crystals

This chapter contains a rather heterogeneous group of crystals. They are
all of low symmetry but it seems not worthwhile, at the present state of
the art, to divide them up into smaller subgroups of (at most) two members.
Useless to say that the lattice dynamics is very difficult but some successful analysis was done for Se and Te [16.7,13]. In the other cases, a
more quantitative investigation is waiting to be carried out.

The following systems are treated:

Crystal	Figures showing		Crystal	Figures showing	
	Dispersion curve $\omega(\underline{q})$	Density of states $D(\omega)$		Dispersion curve $\omega(\underline{q})$	Density of states $D(\omega)$
Se	16.1a	16.1b	D_2O (ice, hex.)	16.8a	16.8b
Te	16.2a	16.2b	α-SiO_2 (quartz)	16.9	
$(SN)_x$	16.3		$MgAl_2O_4$	16.10	
S_8	16.4		K_2SeO_4	16.11	
α-As	16.5a	16.5b	KD_2PO_4	16.12	
NiS	16.6				
Cu_2O	16.7				

Fig. 16.1a. Se: ω(q) [Ref. 16.1, Fig.4], T = RT, M: -, rhombs [16.6],
Lit. [16.2-7]

Fig. 16.1b. Se: D(ω) [Ref. 16.2, Fig.5], M: 9P-VFFM, Lit. [16.3]

<u>Fig. 16.2a.</u> Te: ω(q) [Ref. 16.8, Fig. 3], T = 298 K, M: -, Lit. [16.2,9-13]

<u>Fig. 16.2b.</u> Te: D(ω) [Ref. 16.2, Fig. 6], M: 9P-VFFM, Lit. [16.8,10,12,30]

(SN)$_x$

Fig. 16.3. (SN)$_x$:
$\omega(q)$ [Ref. 16.14,
Fig. 2], T = RT,
M: -, Lit. [16.15]

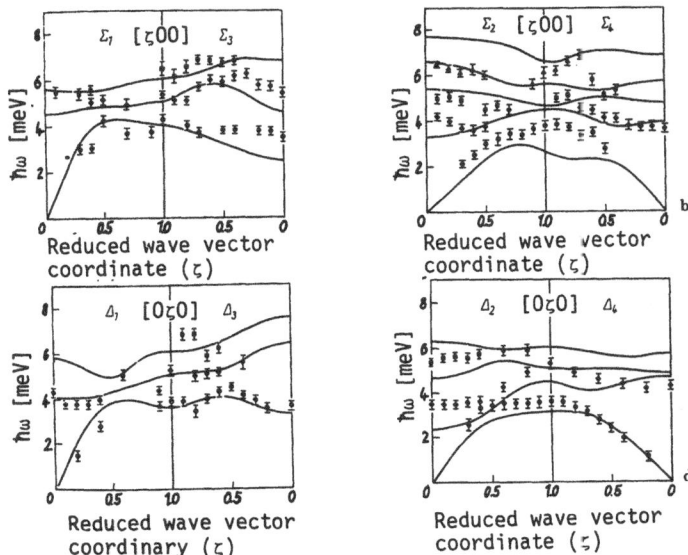

Fig. 16.4. α-S$_8$ ω(q) [Ref. 16.16, Fig. 2], T = RT,
M: 5P-6-exp.-SM, measurements [16.29]

α-As

Fig. 16.5a. α-As: ω($\underline{q}$) [Ref. 16.17, Fig. 4], T = 297 K, M: 27P-FCM

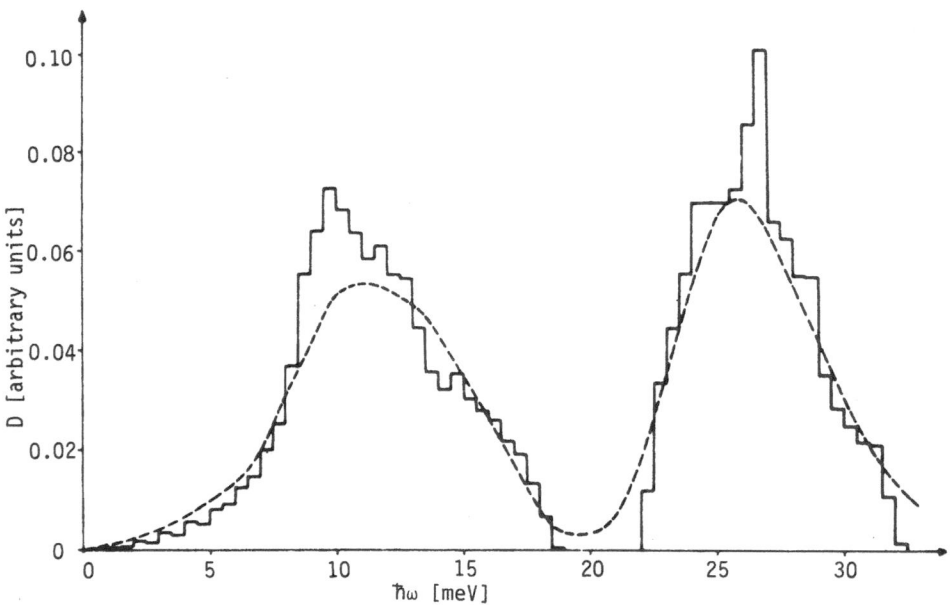

Fig. 16.5b. α-As: D(ω) [Ref. 16.17, Fig. 5], T = 297K, M: 27P-FCM

Fig. 16.6. NiS: ω(q) [Ref.16.18, Fig.3], T = 293 K (full symbols), 44 K (open symbols), M: -

Cu₂O

Fig. 16.7. Cu₂O: ω(q) [Ref.16.19, Fig.3], T = 20°C, M: 3P-RIM (heavy lines), ─── (thin lines), Lit. [16.20]

Fig. 16.8a. Ice (hexagonal) (D$_2$O): ω(q) [Ref. 16.32, Fig. 3], T = 90 K, M: 10P-VFFM, measured points [16.33]

Fig. 16.8b. Ice (hexagonal) (D$_2$O): D(ω) [Ref. 16.32, Fig. 6], T = 90 K, M: 11P-VFFM, Lit. [16.21]

α-SiO₂ (quartz)

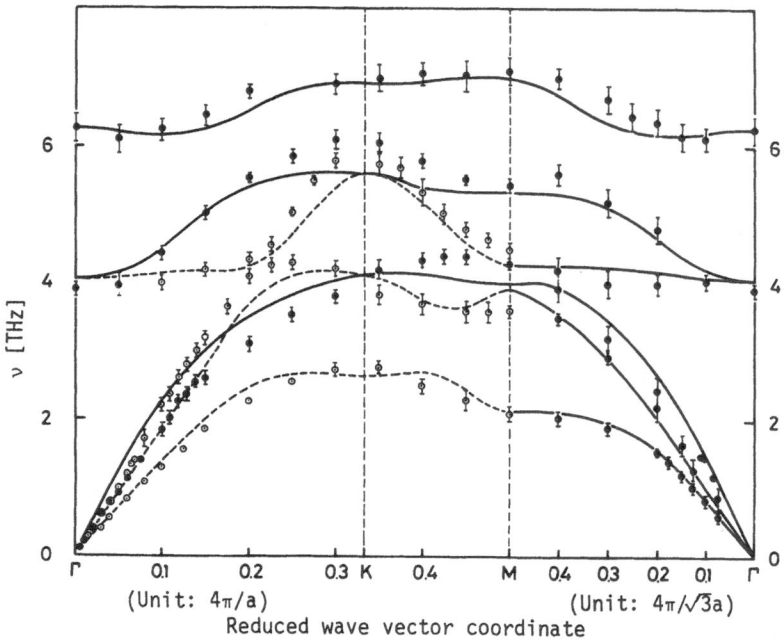

Fig. 16.9. α-quartz (α-SiO$_2$): ω(q) [Ref. 16.23], T = RT, M: 7P-FCM [16.22], Lit. [16.24-26]

Fig. 16.10. MgAl$_2$O$_4$: ω($\underline{q}$) [Ref. 16.27, Fig. 3], T = RT, M: -

K$_2$SeO$_4$

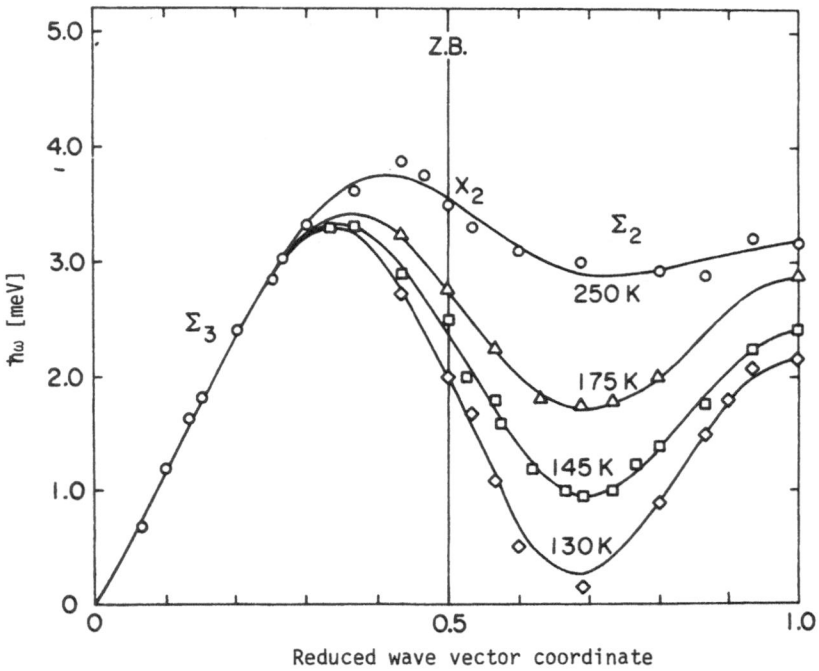

Fig. 16.11. K$_2$SeO$_4$: $\omega(\underline{q})$ [Ref. 16.28, Fig. 9] plotted in an extended zone; Z.B. indicates the zone boundary of the original zone, M: FCM

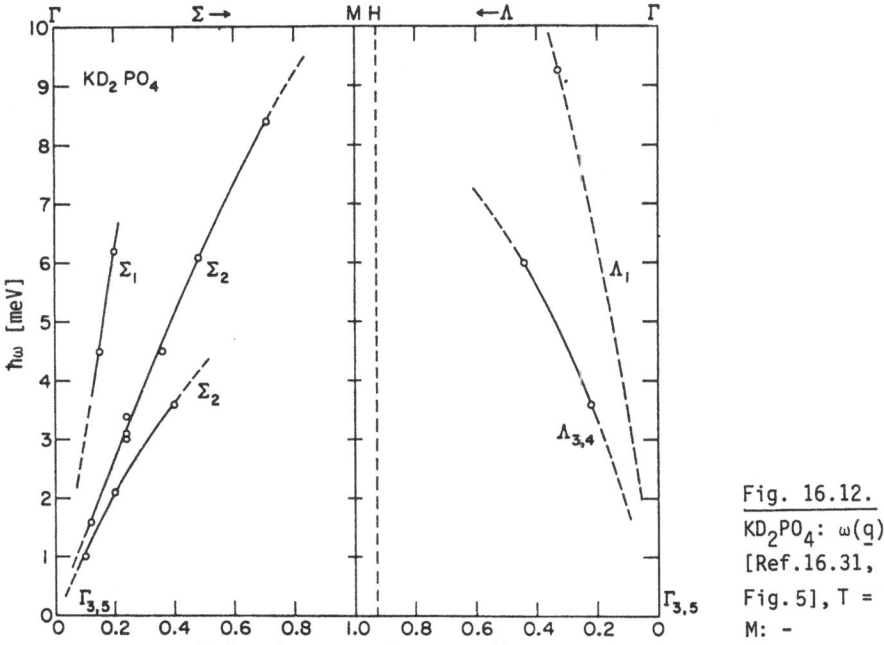

Fig. 16.12.
KD_2PO_4: $\omega(\underline{q})$
[Ref.16.31,
Fig.5], T = RT,
M: -

17. Molecular Crystals

The definition of molecular crystals by the separation of intra- and inter-
molecular vibrations is sometimes difficult. The concept of this separation
is certainly useful and provides a starting point for the theoretical analysis. The field is rapidly growing and more basic treatments are required in
order to understand the delicate playing together of both types of modes in
these crystals. Two molecular crystals, $NaNO_3$ and KNO_3, may be found in
Chap. 14 among the ABO_3 structures.

The following systems are treated:

Crystal	Figures showing	
	Dispersion curve $\omega(\underline{q})$	Density of states $D(\omega)$
α-N_2	17.1	
I_2	17.2	
CO_2	17.3	
NH_3		17.4
D_2O_2	17.5	
NaO_2	17.6	
KN_3	17.7	
SnI_4	17.8	
$NaNO_2$	17.9	

α-N₂

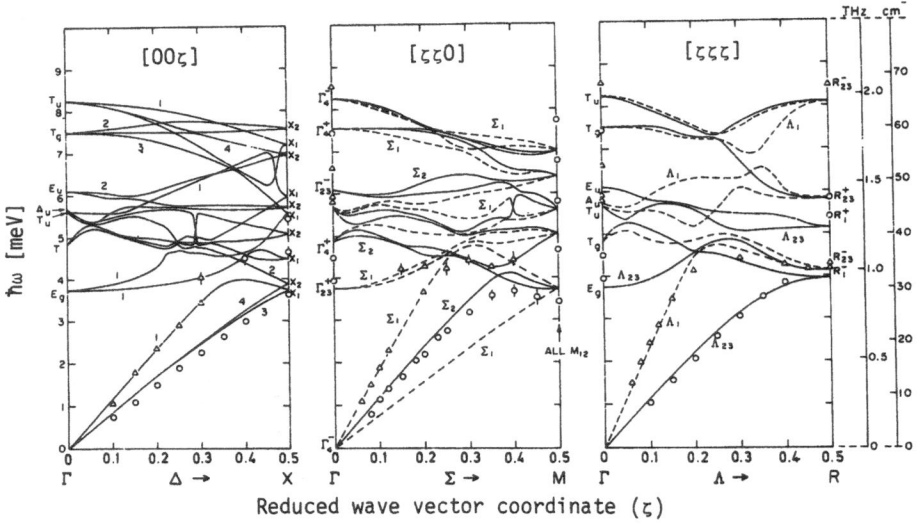

$\hbar\omega$ [meV]

Reduced wave vector coordinate (ζ)

<u>Fig. 17.1.</u> α-N$_2$: ω($\underline{q}$) [Ref. 17.1, Fig. 5], T = 15 K, M: 3P-6/12-pot. including electrostatic quadrupol interaction

ν [THz]

Reduced wave vector coordinate (ζ)

Fig. 17.2. I₂: ω($\underline{q}$) [Ref.17.2, Fig. 3], T = 77 K, measurements [17.3], M: 8P-BCM

CO₂

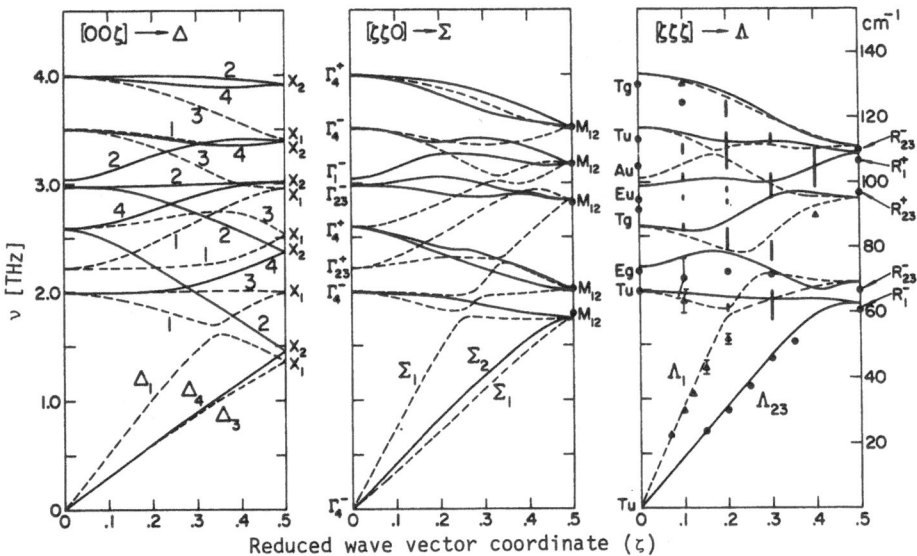

Fig. 17.3. CO_2: $\omega(\underline{q})$ [Ref. 17.4., Fig. 3], T = 95 K, M: 8P-RIM

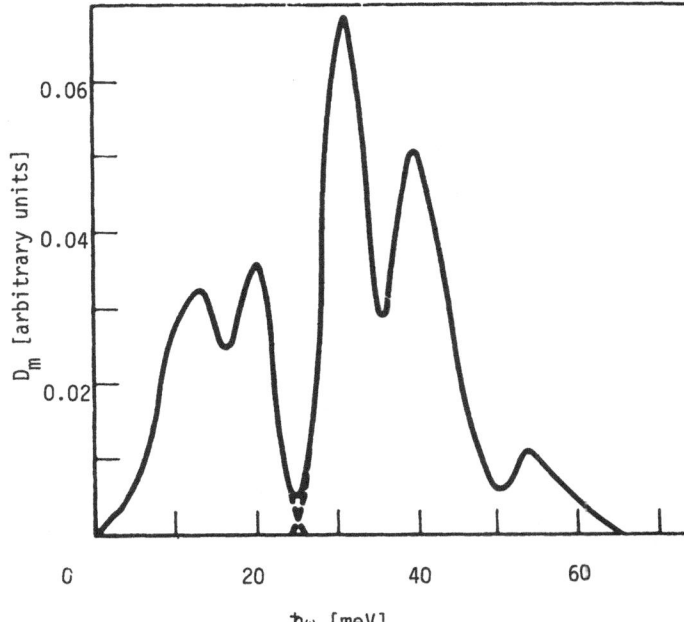

Fig. 17.4. NH₃: D(ω) [Ref. 17.5, Fig. 3], T = 106 K

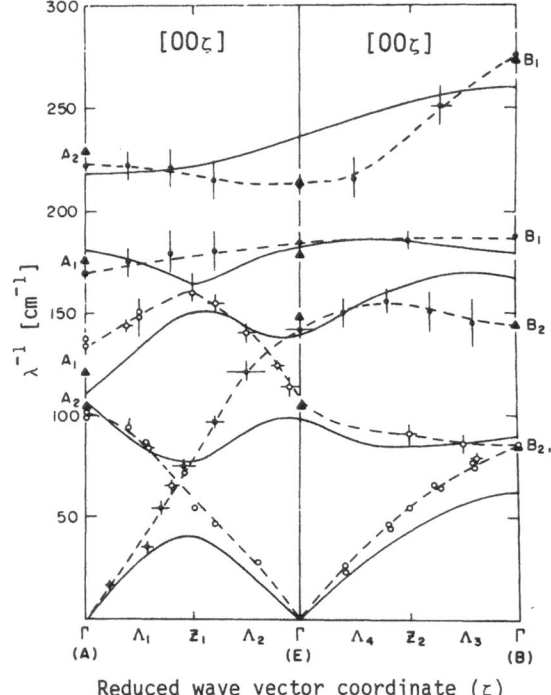

Fig. 17.5. D₂O₂: ω(q) [Ref. 17.6, Fig.1], T = RT, M: VFFM (full lines), dashed curves: guide lines to the eye

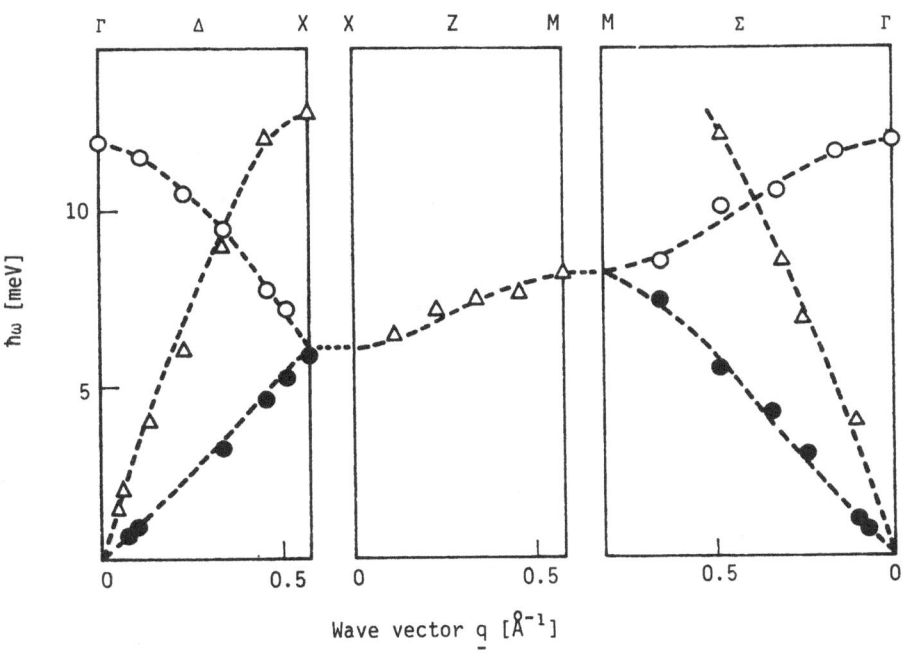

Fig. 17.6. NaO$_2$: $\omega(\underline{q})$ [Ref. 17.7, Fig. 6], T = 225 K, M: –

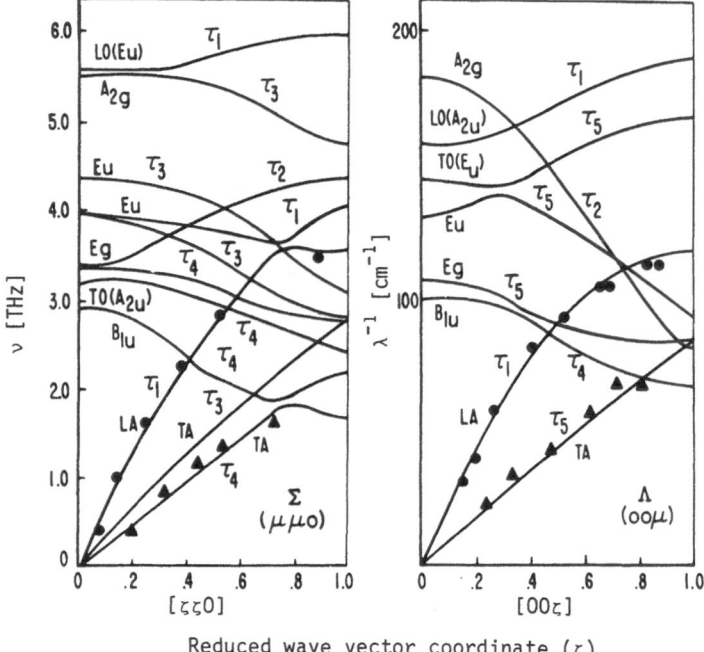

Fig. 17.7. KN$_3$: $\omega(q)$ [Ref. 17.8, Fig. 3], T = RT, M: 2P - pot.-M including Coulomb interactions

Reduced wave vector coordinate (ζ)

Fig. 17.8. SnI$_4$: $\omega(\underline{q})$ [Ref. 17.9, Fig. 3], T = RT, M: −

Fig. 17.9. NaNO$_2$: ω(q) [Ref.17.10, Fig.2], T = 296 K (open points), T = 110 K, (filled points), M: 22P-FCM

18. Mixed Crystals

The few measured mixed crystals are still waiting for a proper theoretical analysis. A tempting idea is the possible correlation to phonons in the perfect crystals by an interpolation scheme which considers the effect of symmetry breaking on high-symmetry phonons such as those at the L or X point in the BZ of rock salt lattice [18.5,8]. Here, the understanding of the linewidth seems of major importance.

The following systems are treated:

Crystal	Figures showing	
	Dispersion curve $\omega(q)$	Density of states $D(\omega)$
$Pb_{0.8}Sn_{0.2}Se$	18.1	
$Pb_{0.87}Sn_{0.13}Te$	18.2	
$NH_4Cl_{1-x}Br_x$	18.3	
$(KBr)_{1-x}(RbBr)_x$	18.4	
$K_{0.5}Rb_{0.5}I$	18.5	

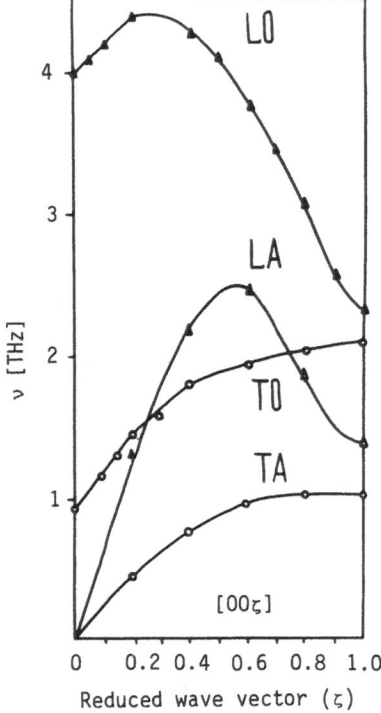

Fig. 18.1. Pb$_{0.8}$ Sn$_{0.2}$Se: $\omega(\underline{q})$ [Ref.18.1, Fig.2], T = 80 K, M: –

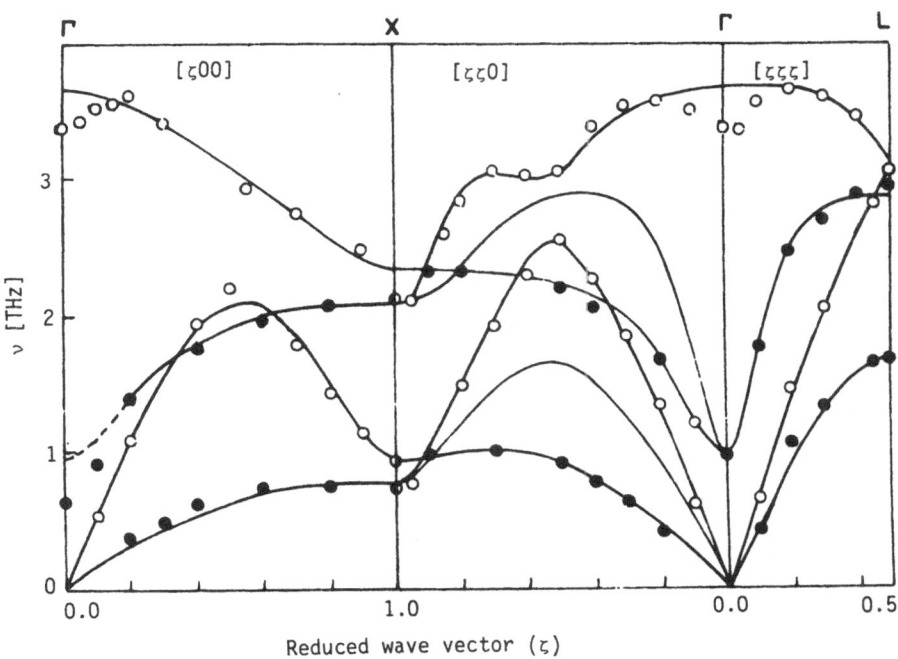

Fig. 18.2. Pb$_{0.87}$ Sn$_{0.13}$ Te: $\omega(\underline{q})$ [Ref. 18.2, Fig.1], T = 82 K, M: 15P-SM

Fig. 18.3. NH$_4$ Cl$_{1-x}$Br$_x$: $\omega(\underline{q})$ [Ref. 18.3, Fig.4], T = RT, | neutron peaks, \\\ neutron bands, ● IR, ◆ Raman, o---o LO [18.4], o-----o TO [18.4], — guide lines to the eye, x = 0 [18.5], x = 1 [18.6]

Reduced wave vector coordinate (ζ)

Reduced wave vector coordinate (ζ)

Fig. 18.4. $(KBr)_{rx}(RbBr)_x$: $\Delta\omega(\underline{q})$ [Ref. 18.7, Fig. 1], T = RT, M: 5P-SM, x = 0.22 crosses, x = 0.45 circles, guide lines to the eye (solid lines), calculations (dashed lines)

Fig. 18.5. K$_{0.5}$Rb$_{0.5}$I: ω($\underline{q}$) [Ref. 18.8, Fig. 3], T = 300 K, M: 11P-SM, Lit. [18.9]

19. Organic Crystals

In the past two decades many organic compounds have been investigated by inelastic neutron scattering. Most of these studies were time of flight measurements of the incoherent scattering from hydrogen atoms in these compounds which yield the density of states weighted by the hydrogen amplitudes. In this book incoherent neutron scattering spectra of organic compounds have not been included since it is extremely difficult to deduce the proper one-phonon density of states from these spectra.

Phonon dispersion curves have been determined by coherent inelastic neutron scattering only in a few organic compounds due to difficulties in growing suitable single crystals. The theoretical description of the measured dispersion curves uses mainly sets of formal force constants, but there is no doubt that this area will attract more sophisticated theoretical work in the near future.

The following systems are treated:

| | | Figures showing | |
		Dispersion curve $\omega(q)$	Density of states $D(\omega)$
CD_4	(deutero-) methane	19.1a,b	
$C_4N_2D_4$	pyrazine	19.2	
$C_6D_4Cl_2$	p-dichloro-benzene	19.3	
$C_6D_{12}N_4$	hexa-methylen-tetramine	19.4a	19.4b
$C_5D_6O_2N_2$	methyl-thymine	19.5	
$C_4D_5ON_3$	cytosine (-hydrate)	19.6	
$C_{10}D_8$	naphtalene	19.7a	
$C_{14}D_{10}$	anthracene	19.8	
$CO(ND_2)_2$	urea	19.9	

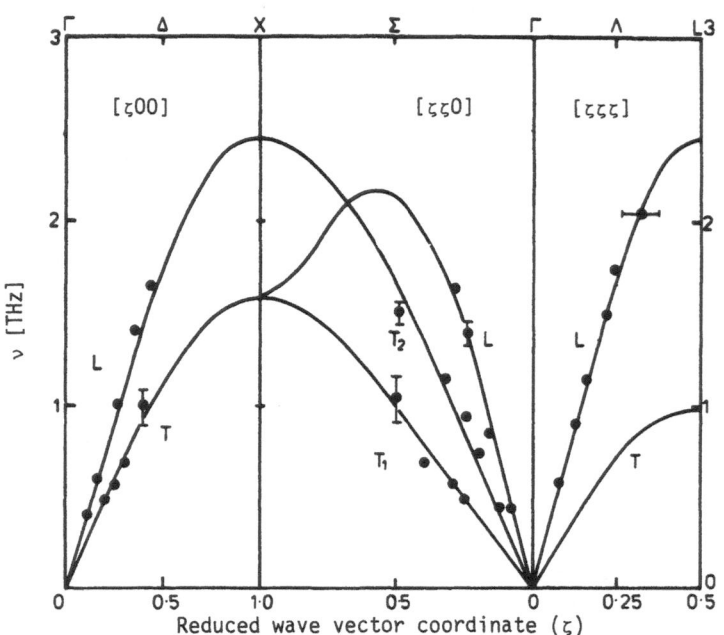

Fig. 19.1a. Methane (CD$_4$): ω(q̠) [Ref. 19.1, Fig. 5], T = 32,5 K (phase I),
M: 3P-FCM, Lit. [19.2-5]

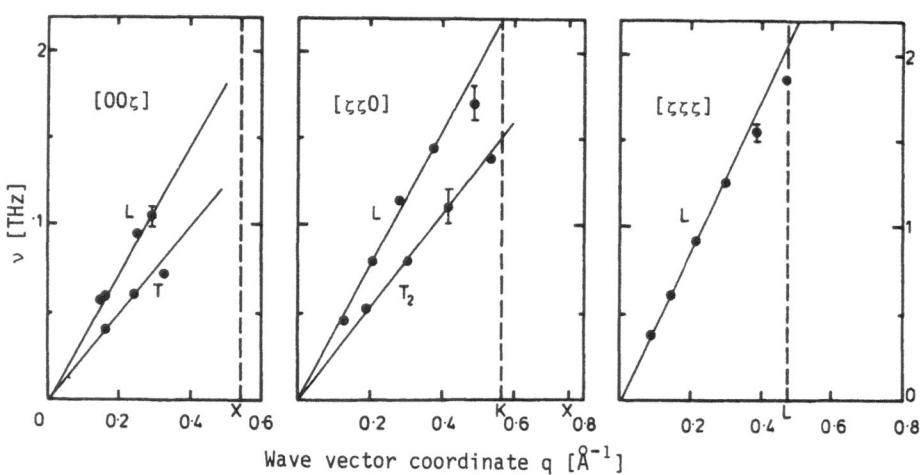

Fig. 19.1b. Methane (CD$_4$): ω(q̠) [Ref. 19.1, Fig. 4], T = 22.8 K (phase II)

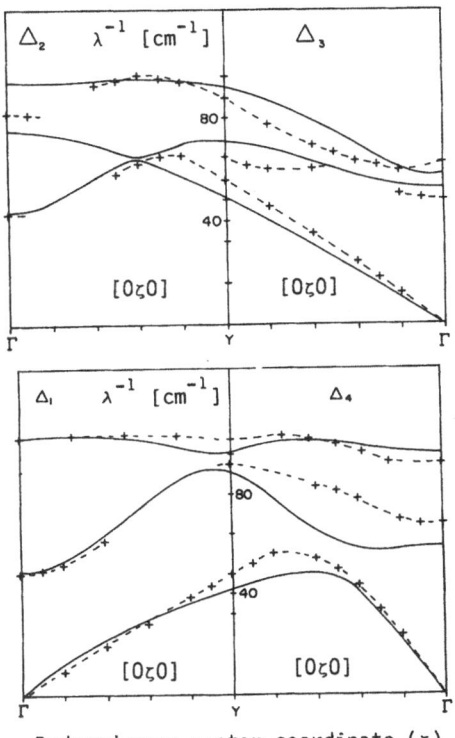

Reduced wave vector coordinate (ζ)

Fig. 19.2. Pyrazine ($C_4N_2D_4$): $\omega(\underline{q})$
[Ref. 19.6, Fig.3], T = 295 K,
M: 19P-6-exp.-pot.-M (including
electrical dipole moment)

Fig. 19.3. p-dichloro-
benzene (C$_6$D$_4$Cl$_2$): $\omega(\underline{q})$
[Ref. 19.7, Fig. 4],
T = 295 K (a), T = 90 K (b),
M: -

Fig. 19.4a. Hexamethylen-tetramine (C₆D₁₂N₄): ω(q) [Ref. 19.8, Fig.2],
T = 100 K, M: 5P-FCM (solid and dashed lines), 4P-FCM (dotted lines),
Lit. [19.9-12]

Fig. 19.4b. Hexamethylen-tetramine (HMT): D(ω) [Ref. 19.8, Fig. 5], T = 100K, M: 5P-FCM, HMT (solid lines), deuterated HMT (dashed lines)

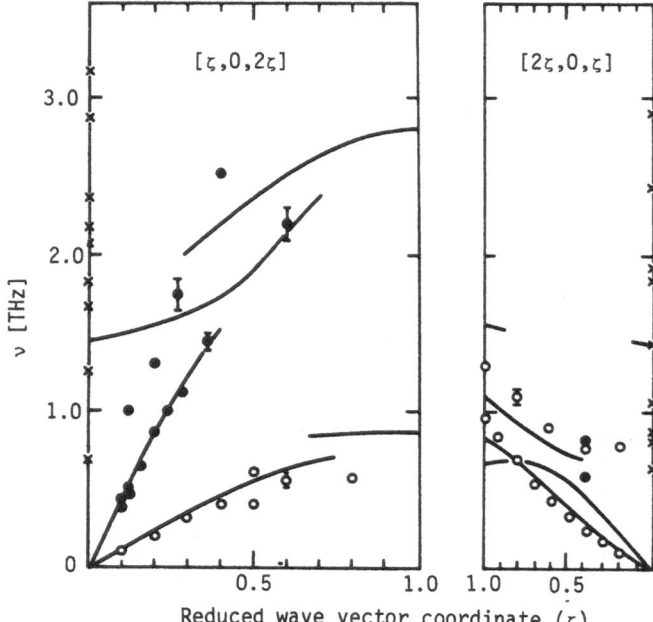

Fig. 19.5.
1-methylthymine
($C_5D_6O_2N_2$): $\omega(\underline{q})$
[Ref. 19.13,
Fig. 2], T = 296 K,
M: 28P-ASM,
Lit. [19.14]

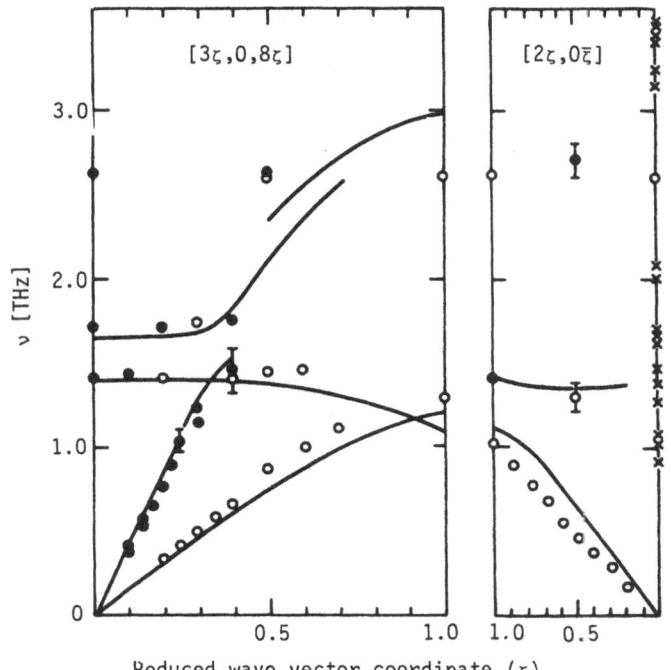

Fig. 19.6.
cytosine mono-
hydrate ($C_4D_5ON_3$):
$\omega(\underline{q})$ [Ref. 19.13,
Fig. 3], T = 296 K,
M: 28P-ASM

Fig. 19.7a. Naphthalene ($C_{10}D_8$): $\omega(q)$ [Ref. 19.15, Fig. 2a], T = 98 K, 5 K, M: -, Lit. [19.16-21]

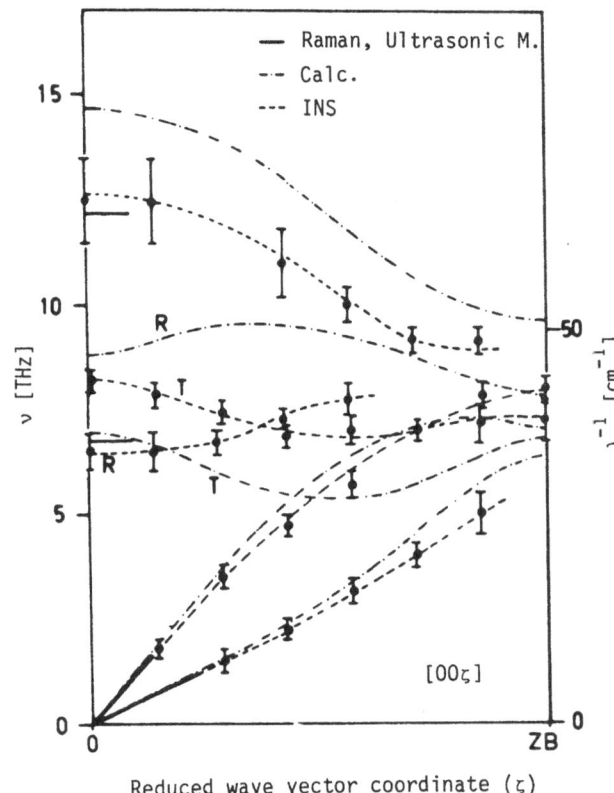

Fig. 19.8.
Anthracene ($C_{14}D_{10}$):
$\omega(\underline{q})$ [Ref. 19.22,
Fig.1], T = RT,
M: 7P-FCM

Fig. 19.9. Urea (CO (ND$_2$)$_2$): $\omega(\underline{q})$ [Ref. 19.23, Fig. 4], T = RT, M: 12P-FCM
(including long range dipole-dipole interactions)

References

2.1   H. Bilz, D. Strauch, R. Wehner: *Handbuch für Physik*, Vol. 25/2d
      (Springer Berlin, Heidelberg, New York 1979)
2.2   J.I. Birman: *Handbuch für Physik*, Vol. 25/2b (Springer Berlin, Heidel-
      berg, New York 1974)
2.3   M. Born, K. Huang: *Dynamical Theory of Crystal Lattices* (Oxford Uni-
      versity Press, London 1954)
2.4   G. Leibfried, W. Ludwig: Solid State Phys. *12*, 275 (1961)
2.5   R.A. Cowley: Phil. Mag. *11*, 673 (1965)
2.6   R.S. Leigh, B. Szigeti, V.K. Tewary: Proc. Roy. Soc. A *320*, 505 (1971)
2.7   A.A. Maradudin, E.W. Montroll, G.H. Weiss, I.P. Ipatova: Solid State
      Phys. Suppl. *3*, 1 (1972)
2.8   W. Cochran: Crit. Rev. Solid State Sci. *2*, 1 (1971)
      J. Hardy in *Dynamical Properties of Solids*, Vol. 1, ed. by G.K. Horton
2.9   A.A. Maradudin (North Holland, Amsterdam 1974) p. 157
2.10 S.K. Sinha: Crit. Rev. Solid State Sci. *3*, 273 (1973)
2.11 H. Bilz, B. Gliss, W. Hanke in *Dynamical Properties of Solids*, Vol. 1,
      ed. by G.K. Horton, A.A. Maradudin (North Holland, Amsterdam 1974) p. 343
2.12 W. Hanke: Adv. in Phys. *27*, 287 (1978)
2.13 R.A. Cowley, W. Cochran, B.N. Brockhouse, A.D.B. Woods: Phys. Rev. *131*,
      1030 (1963)
2.14 H. Bilz, M. Buchanan, K. Fischer, R. Haberkorn: Solid State Commun.
      16, 1023 (1975)
2.15 K. Kunc, H. Bilz: Solid State Commun. *19*, 1027 (1976)
2.16 S.S. Jaswal: Phys. Rev. Lett. *35*, 1600 (1975)
2.17 R. Zeyher: Phys. Rev. Lett. *35*, 174 (1975)
2.18 U. Schröder: Solid State Commun. *4*, 347 (1966)
2.19 P. Allen: Phys. Rev. B *16*, 5139 (1977)
2.20 W. Kress, H. Bilz: to be published
2.21 W. Weber: Phys. Rev. B *8*, 5082 (1973)
2.22 A.N. Basu, D. Roy, S. Sengupta: Phys. Status Solidi (a) *23*, 11 (1974)
2.23 K. Fischer, H. Bilz, R. Haberkorn, W. Weber: Phys. Status Solidi (b)
      54, 285 (1972)
2.24 B. Dorner, W. von der Osten, W. Bührer: J. Phys. C *9*, 723 (1976)
2.25 W. Kleppmann, W. Weber: to be published
2.26 D.H. Kühner, H.V. Lauer, W.E. Bron: Phys. Rev. B *5*, 4112 (1972)
2.27 K.B. Tolpygo: Ukr. Fiz. Zh *18*, 1636 (1973);
      O.N. Bolonin, K.B. Tolpygo: Sov. Phys. Solid State *18*, 446 (1976)
2.28 W. Cochran: Proc. Roy. Soc. (London) A *253*, 260 (1959)
2.29 W. Kress: Phys. Status Solidi (b) *49*, 235 (1972)
2.30 W. Weber: Phys. Rev. B *15*, 4789 (1977)
2.31 P.N. Keating: Phys. Rev. *140* A, 369 (1965)
2.32 E.B. Wilson, jr., J.C. Decius, P.C. Cross: *Molecular Vibrations*
      (McGraw-Hill, New York 1955)
2.33 J.H. Schachtschneider, R.G. Snyder: Spectrochim. Acta *19*, 117 (1963)
2.34 R. Tubino, L. Piseri, G. Zerbi: J. Chem. Phys. *6*, 1022 (1972)
2.35 K. Kunc, M. Balkanski, M. Nusimovici: Phys. Rev. B *12*, 4346 (1975)

2.36 R.K. Singh, M.P. Verma: Phys. Status Solidi *36*, 335 (1969),
      38, 851 (1970)
2.37 R. Zeyher: Phys. Status Solidi (b) *48*, 711 (1971)
2.38 R.K. Singh, H.N. Gupta, M.K. Agrawal: Phys. Rev. B *17*, 894 (1978)
2.39 G.K. Horton, A.A. Maradudin (Eds.): *Dynamical Properties of Solids*,
      Vol. 1 (North Holland Amsterdam 1974)
2.40 R. Zeyher: Phys. Rev. Lett. *35*, 174 (1975)
3.1   C. Stassis, G. Kline, W. Kamitakara, S.K. Sinha: Phys. Rev. B *17*, 1130
      (1978)
3.2   H. Horner: Solid State Commun. *9*, 79 (1971)
3.3   R.A. Reese, S.K. Sinha, T.O. Brun, C.R. Tilford: Phys. Rev. A *3*, 1688
      (1971)
3.4   V.J. Minkiewicz, T.A. Kitchens, F.P. Lipschultz, R. Nathans,
      G. Shirane: Phys. Rev. *174*, 267 (1968)
3.5   V.J. Minkiewicz, T.A. Kitchens, G. Shirane, E.B. Osgood: Phys. Rev.
      A *8*, 1513 (1973)
3.6   H.R. Glyde: J. Low. Temp. Phys. *3*, 559 (1970)
3.7   H.R. Glyde: Can. J. Phys. *49*, 761 (1971)
3.8   J.A. Leake, W.B. Daniels, J. Skalyo, Jr., B.C. Frazer, G. Shirane:
      Phys. Rev. *181*, 1251 (1969)
3.9   J. Skalyo, Jr., V.J. Minkiewicz, G. Shirane, W.B. Daniels: Phys. Rev.
      B *6*, 4766 (1972)
3.10 J. Eckert, W.B. Daniels, J.D. Axe: Phys. Rev. B *14*, 3649 (1976)
3.11 Y. Endoh, G. Shirane, J. Skalyo, Jr.: Phys. Rev. B *11*, 1681 (1975)
3.12 Y. Fujii, N.A. Lurie, R. Pynn, G. Shirane: Phys. Rev. B *10*, 3647
      (1974)
3.13 J.A. Barker, M.L. Klein, M.V. Bobetic: Phys. Rev. B *2*, 4176 (1970)
3.14 J. Skalyo, Jr., Y. Endoh, G. Shirane: Phys. Rev. B *9*, 1797 (1974)
3.15 J.A. Barker, M.V. Bobetic, M.L. Klein: Phys. Lett. A *34*, 415 (1971)
3.16 N.A. Lurie, G. Shirane, J. Skalyo, Jr.: Phys. Rev. B *9*, 5300 (1974)
3.17 B.J. Palmer, D.H. Saunderson, D.N. Batchelder: J. Phys. C *6*, L313
      (1973)
4.1   J.L. Verble, J.L. Warren, J.L. Yarnell: Phys. Rev. *168*, 980 (1968)
4.2   M.P. Verma, R.K. Singh: J. Phys. C *4*, 2749 (1971)
4.3   R.K. Singh: J. Phys. C *7*, 3473 (1974)
4.4   R. Zeyher: Phys. Rev. Lett. *35*, 174 (1975)
4.5   R. Zeyher in Proc. Conf. Neutron Scattering, Vol. 1 (Gatlinburg,
      USA 1976) p. 550
4.6   S.S. Jaswal, V.D. Dilly: Phys. Rev. B *15*, 2366 (1977)
4.7   D. Laplace: J. Phys. C *10*, 3499 (1977)
4.8   R. Zeyher in *Lattice Dynamics*, ed. by M. Balkanski (Flammarion,
      Paris 1978) p.17
4.9   M.G. Zemlianov, E.G. Brovman, N.A. Chernoplekov, Iu.L. Shitikov in
      Inelastic Scattering of Neutrons, Vol. 2 (IAEA, Vienna 1965) p. 431
4.10 G. Dolling, H.G. Smith, R.M. Nicklow, P.R. Vijayaraghavan,
      M.K. Wilkinson: Phys. Rev. *168*, 970 (1968)
4.11 V. Nüßlein, U. Schröder: Phys. Status Solidi *21*, 309 (1967)
4.12 A.N. Basu, S. Sengupta: J. Phys. C *5*, 1158 (1972)
4.13 A. Rastogi, J.P. Hawranek, R.P. Lowndes: Phys. Rev. B *9*, 1938 (1974)
4.14 R.K. Singh: J. Phys. C *7*, 3473 (1974)
4.15 A.D.B. Woods, B.N. Brockhouse, M. Sakomoto, R.N. Sinclair: *Inelastic
      Scattering of Neutrons in Solids and Liquids*, Vol. 1 (IAEA, Vienna
      1961) p. 487
4.16 W.J.L. Buyers: Phys. Rev. *153*, 923 (1967)
4.17 I.S. Braude: Sov. Phys. Solid State *13*, 327 (1971)
4.18 J.S. Melvin, J.D. Pirie, T. Smith: Phys. Rev. *175*, 1082 (1968)
4.19 A.M. Karo, J.R. Hardy: Phys. Rev. *181*, 1272 (1969)
4.20 R.K. Singh, M.P. Verma: Phys. Rev. B *2*, 4289 (1970)

4.21 K.V. Namjoshi, S.S. Mitra, J.F. Vetelino: Solid State Commun. *9*, 185 (1971)
4.22 A. Ghosh, A.N. Basu, S. Sengupta: Proc. Roy. Soc. (GB) A *340*, 199 (1974)
4.23 Sneh, B. Dayal: Phys. Status Solidi (b) *67*, 125 (1975)
4.24 R.K. Singh, K. Chandra: Phys. Rev. B *14*, 2625 (1976)
4.25 R.W. MacPherson, T. Timusk: Can. J. Phys. *48*, 2917 (1970)
4.26 G. Raunio, S. Rolandson: Phys. Rev. B *2*, 2098 (1970)
4.27 G. Raunio, L. Almquist, R. Stedman: Phys. Rev. *178*, 1496 (1969)
4.28 R.E. Schmunk, D.R. Winder: J. Phys. Chem. Solids *31*, 131 (1970)
4.29 E.R. Cowley: J. Phys. C *5*, 1345 (1972)
4.30 O.N. Bolonin: Sov. Phys. Solid State *18*, 1415 (1976)
4.31 O.N. Bolonin, K.B. Tolpygo: Sov. Phys. Solid State *18*, 446 (1976)
4.32 S.K. Sarkar, S. Sengupta: Phys. Status Solidi (b) *87*, 517 (1978)
4.33 A.N. Basu, S. Sengupta in *Lattice Dynamics*, ed. by M. Balkanski (Flammarion, Paris 1978) p.43
4.34 J.S. Reid, T. Smith, W.J.L. Buyers: Phys. Rev. B *1*, 1833 (1970)
4.35 O.N. Bolonin: Sov. Phys. Solid State *19*, 1088 (1977)
4.36 U. Schröder: Solid State Commun. *4*, 347 (1966)
4.37 A.D.B. Woods, B.N. Brockhouse, R.A. Cowley, W. Cochran: Phys. Rev. *131*, 1025 (1963)
4.38 A.D.B. Woods, W. Cochran, B.N. Brockhouse: Phys. Rev. *119*, 980 (1960)
4.39 R.A. Cowley, W. Cochran, B.N. Brockhouse, A.D.B. Woods: Phys. Rev. *131*, 1030 (1963)
4.40 A.N. Basu, S. Sengupta: Phys. Status Solidi *29*, 367 (1968)
4.41 T.I. Kucher, O.E. Tomasevich: Sov. Phys. Solid State *12*, 423 (1970)
4.42 I.S. Braude: Sov. Phys. Solid State *13*, 327 (1971)
4.43 A. Loidl, H. Jex, J. Daubert, M. Müllner: Phys. Status Solidi (b) *76*, 581 (1976)
4.44 D. Bäuerle, B. Fritz: Phys. Status Solidi *24*, 207 (1967)
4.45 W. Bührer: Phys. Status Solidi *41*, 789 (1970)
4.46 R.K. Singh, M.P. Verma: Phys. Status Solidi *38*, 851 (1970)
4.47 J.R.D. Copley, R.W. MacPherson, T. Timusk: Phys. Rev. *182*, 965 (1969)
4.48 G. Raunio, L. Almquist: Phys. Status Solidi *33*, 209 (1969)
4.49 N. Wakabayashi, S.K. Sinha: Phys. Rev. B *10*, 745 (1974)
4.50 A.N. Basu, S. Sengupta: Phys. Rev. B *14*, 2635 (1976)
4.51 R.A. Cowley, W. Cochran, B.N. Brockhouse, A.D.B. Woods: Phys. Rev. *131*, 1030 (1963)
4.52 E.C. Svenson, W.L.J. Buyers: Phys. Rev. *165*, 1063 (1968)
4.53 R.A. Cowley, E.C. Svenson, W.L.J. Buyers: Phys. Rev. Lett. *23*, 525 (1969)
4.54 W. Bluthardt, W. Schneider, M. Wagner: Phys. Status Solidi (b) *56*, 453 (1973)
4.55 G. Dolling, R.A. Cowley, C. Schittenhelm, J.M. Thorson: Phys. Rev. *147*, 577 (1966)
4.56 D. Bäuerle, R. Hübner: Phys. Rev. B *2*, 4252 (1970)
4.57 G. Raunio, S. Rolandson: J. Phys. C *3*, 1013 (1970)
4.58 H.H. Lal, M.P. Verma: J. Phys. C *5*, 543 (1972)
4.59 M.J.L. Sangster, R.M. Atwood: J. Phys. C *11*, 1541 (1978)
4.60 S. Rolandson, G. Raunio: J. Phys. C *4*, 958 (1971)
4.61 G. Raunio, S. Rolandson: Phys. Status Solidi *40*, 749 (1970)
4.62 W. Kress: Phys. Status Solidi (b) *62*, 403 (1974)
4.63 K.V. Namjoshi, S.S. Mitra, J.F. Vetelino in *Phonons*, ed. by M.A. Nusimovici (Flammarion, Paris 1971) p.79
4.64 K.V. Namjoshi, S.S. Mitra, J.F. Vetelino, Phys. Rev. B *3*, 4398 (1971)
4.65 R.K. Singh, H.N. Gupta, M.K. Agrawal: Phys. Rev. B *17*, 894 (1978)
5.1   M.J.L. Sangster, G. Peckham, D.H. Saunderson: J. Phys. C *3*, 1026 (1970)

5.2   K.V. Namjoshi, S.S. Mitra, J.F. Vetelino: Solid State Commun. *9*, 185
      (1971)
5.3   N.S. Gillis: Phys. Rev. B *3*, 1482 (1971)
5.4   M. Caltier, A. Montaner in *Phonons*, ed. by M.A. Nusimovici,
      (Flammarion, Paris 1971) p.59
5.5   R.K. Singh, K.S. Upadhyaya: Phys. Rev. B *6*, 1589 (1972)
5.6   A. Ghosh, A.N. Basu in *Lattice Dynamics*, ed. by M. Balkanski
      (Flammarion, Paris 1978) p.65
5.7   K.H. Rieder, B.A. Weinstein, M. Cardona, H. Bilz: Phys. Rev. B *8*,
      4780 (1973)
5.8   D.H. Saunderson, G. Peckham: J. Phys. C *4*, 2009 (1971)
5.9   R.P. Vijayaraghavan, Marsongkohari, P.K. Iyengar in *Neutron Inelastic
      Scattering* (IAEA, Vienna 1972) p.95
5.10 K.S. Upadhyaya, R.K. Singh: J. Phys. Chem. Solids *36*, 293 (1975)
5.11 K.H. Rieder, R. Migoni, B. Renker: Phys. Rev. B *12*, 3374 (1975)
5.12 S.K. Agrawal: J. Phys. Chem. Solids *38*, 199 (1977)
5.13 S.S. Chang, C.W. Tompson, E. Gürmen, L.D. Mühlestein: J. Phys. Chem.
      Solids *36*, 769 (1975)
5.14 W. Kress, W. Reichardt, V. Wagner, G. Kugel, B. Hennion in *Lattice
      Dynamics*, ed. by M. Balkanski (Flammarion, Paris 1978) p.77
5.15 B.C. Haywood, M.F. Collins: J. Phys. C *4*, 1299 (1971)
5.16 K.S. Upadhyaya, R.K. Singh: J. Phys. Chem. Solids *35*, 1175 (1974)
5.17 V. Wagner, W. Reichardt, W. Kress: *Proc. Conf. Neutron Scattering*,
      Vol. 1 (Gatlinburg, USA 1976) p.175
5.18 B.R.K. Gupta, M.P. Verma: J. Phys. Chem. Solids *38*, 929 (1977)
5.19 G. Kugel, C. Carabatos, B. Hennion, B. Prevot, A. Revcolevschi,
      D. Tocchetti: Phys. Rev. B *16*, 378 (1977)
5.20 J. Sakurai, W.J.S. Buyers, R.A. Cowley, G. Dolling: Phys. Rev. *167*
      510 (1968)
5.21 W.J.L. Buyers, G. Dolling, J. Sakurai, R.A. Cowley in *Neutron Inelastic
      Scattering*, Vol. 2 (IAEA Vienna 1968) p.123
5.22 W. Reichardt, V. Wagner, W. Kress: J. Phys. C *8*, 3955 (1975)
5.23 R.A. Coy, C.W. Tompson, E. Gürmen: Solid State Commun. *18*, 845 (1976)
5.24 R.P. Goyal: Phys. Status Solidi (b) *79*, K115 (1977)
5.25 M.P. Verma, S.K. Agraval: Phys. Rev. B *8*, 4880 (1973)
5.26 T.S. Chen, F.W. DeWette, L. Kleinman, D.G. Dempey: Phys. Rev. B *17*,
      844 (1978)
6.1   P. Roedhammer, W. Reichardt, F. Holtzberg: Phys. Rev. Lett. *40*, 465
      (1978)
6.2   P. Roedhammer, W. Reichardt, F. Holtzberg in *Lattice Dynamics*, ed. by
      M. Balkanski (Flammarion, Paris 1978) p.84
6.3   L. Pintschovius, W. Reichardt, B. Scheerer: J. Phys. C *11*, 1557 (1978)
6.4   W. Kress, P. Roedhammer, H. Bilz, W.D. Teuchert, A.N. Christensen:
      Phys. Rev. B *17*, 111 (1978)
6.5   K.H. Rieder, W. Drexel: Phys. Rev. Lett. *34*, 148 (1975)
6.6   W. Weber: Phys. Rev. B *8*, 5082 (1973)
6.7   H.G. Smith in *Superconductivity in d and f Band Transition Metals*,
      ed. by D.H. Douglass (AIP, New York 1972) p.321
6.8   H.G. Smith, N. Wakabayashi, R.M. Nicklow, S. Mihailovich in: *Proc.
      13th Int. Conf. on Low Temperature Physics*, ed. by R.H. Kropschot,
      K.P. Timmerhaus (Boulder, USA 1972)
6.9   B.R.K. Gupta, M.P. Verma: J. Phys. Chem. Solids *37*, 815 (1976)
6.10 F. Gompf, J. Salgado, W. Reichardt: KFK-Report 2054 (1973/74) p.21
6.11 F. Gompf: KFK-Report 2357 (1975/76) p.9
6.12 H.G. Smith, W. Gläser: Phys. Rev. Lett. *25*, 1611 (1970)
6.13 W. Weber, H. Bilz, U. Schröder: Phys. Rev. Lett. *28*, 600 (1972)
6.14 M. Mostoller: Phys. Rev. B *5*, 1260 (1972)
6.15 M.P. Verma, B.R.K. Gupta: Phys. Rev. B *12*, 1314 (1975)

6.16 L. Pintschovius, W. Reichardt, B. Scheerer: KFK-Report 2538 (1977) p.4
6.17 H.G. Smith, W. Gläser in *Phonons*, ed. by M.A. Nusimovici (Flammarion, Paris 1971)
6.18 F. Gompf, L. Pintschovius, W. Reichardt, B. Scheerer in: *Proc. Conf. Neutron Scattering*, Vol. 1, ed. by R.M. Moon (Gatlinburg, USA 1976) CONF-760601, p.129
6.19 W. Hanke, J. Hafner, H. Bilz: Phys. Rev. Lett. *37*, 1560 (1976)
6.20 B. Splettstösser: Z. Phys. B *26*, 151 (1977)
6.21 A.K. Goyal, S.C. Goyal: Phys. Status Solidi (b) *81*, K63 (1977)
6.22 F. Gompf, W. Reichardt, A.N. Christensen: KFK-Report 2357 (1975/76) p.11
6.23 S.C. Goyal, A.K. Goyal: Phys. Status Solidi (b) *77*, 639 (1976)
6.24 G. Dolling, T.M. Holden, E.C. Svensson, W.J.L. Buyers, G.H. Lander in *Lattice Dynamics*, ed. by M. Balkanski (Flammarion, Paris 1978) p.81
6.25 V. Varma, W. Weber: Phys. Rev. Lett. *39*, 1094 (1977)
6.26 F. Gompf, W. Reichardt: KFK-Report 2538 (1977) p.6
7.1   E.R. Cowley, J.K. Darby, G.S. Pawley: J. Phys. C *2*, 1916 (1969)
7.2   G.S. Pawley, W. Cochran, R.A. Cowley, G. Dolling: Phys. Rev. Lett. *17*, 753 (1966)
7.3   M.M. Elcombe: Proc. Roy. Soc. (London) A *300*, 210 (1967)
7.4   P.R. Vijayaraghavan, S.K. Sinha, P.K. Iyengar in *Proc. Nuc. Phys. and Solid State Symp.* (Bangalore, India 1973) p.208
7.5   W. Cochran, R.A. Cowley, G. Dolling, M.M. Elcombe: Proc. Roy. Soc. (London) A *293*, 433 (1966)
7.6   H.A. Alperin, S.J. Pickart, J.J. Rhyne, V.J. Minkiewicz: Phys. Lett. A *40*, 295 (1972)
7.7   N. Wakabayashi, A. Furrer: Phys. Rev. B *13*, 4343 (1976)
7.8   P.R. Vijayaraghavan, R.M. Nicklow, H.G. Smith, M.K. Wilkinson: Phys. Rev. B *1*, 4819 (1970)
7.9   R.K. Singh, K.S. Upadhyaya: Phys. Status Solidi (b) *51*, 389 (1972)
7.10 K. Fischer, H. Bilz, R. Haberkorn, W. Weber: Phys. Status Solidi (b) *54*, 285 (1972)
7.11 K.S. Upadhyaya, P.S. Mahesh: Phys. Status Solidi (b) *59*, 279 (1973)
7.12 J.P. Hawranek, R.P. Lowndes: Solid State Commun. *11*, 1473 (1972)
7.13 H. Kanzaki, S. Sakurayi, S. Hoshino, J. Shirane, Y. Fujii: Solid State Commun. *15*, 1547 (1974)
7.14 B. Dorner, W. von der Osten, W. Bührer: J. Phys. C *9*, 723 (1976)
7.15 W. von der Osten, B. Dorner: Solid State Commun. *16*, 431 (1975)
7.16 W. Bührer: Phys. Status Solidi (b) *66*, 739 (1975)
7.17 R.K. Singh, K.P. Singh: Phys. Status Solidi (b) *78*, 677 (1976)
7.18 Y. Fujii, S. Hoshino, S. Sakurayi, H. Kanzaki, J.W. Lynn, G. Shirane: Phys. Rev. B *15*, 358 (1977)
7.19 B. Dorner, J. Windscheif, W. von der Osten in: *Lattice Dynamics*, ed. by M. Balkanski, (Flammarion, Paris 1978) p.535
7.20 R. Zeyher in: *Lattice Dynamics*, ed. by M. Balkanski (Flammarion, Paris 1978) p.17
7.21 J.M. Rowe, J.J. Rush, N. Vagelatos, D.L. Price, D.G. Hinks, S. Susman: J. Chem. Phys. *62*, 4551 (1975)
7.22 C.J. Bill, H. Jex, M. Müllner: Phys. Lett. A *56*, 320 (1976)
7.23 H. Jex, C.J. Maetz: Phys. Status Solidi (b) *85*, 511 (1978)
7.24 J. Daubert, K. Knorr, W. Dultz, H. Jex, R. Currat: J. Phys. C *9*, L389 (1976)
7.25 D. Fontaine, H. Poulet: Phys. Status Solidi (b) *58* K9 (1973)
7.26 W. Dultz: Solid State Commun. *15*, 595 (1974)
7.27 N. Nücker, K. Knorr, H. Jex: J. Phys. C *11*, ·1 (1977)
7.28 N. Vagelatos, J.M. Rowe, J.J. Rush: Phys. Rev. B *12*, 4522 (1975)

7.29   C.M. Goel. T.P. Sharma, B. Dayal: J. Phys. Chem. Solids *38*, 1285 (1977)
7.30   S. Rolandson: Phys. Status Solids (b) *52*, 643 (1972)
7.31   W. Bühner: J. Phys. C *6*, 2931 (1973)
7.32   H.N. Gupta, R.K. Singh: Phys. Status Solidi (b) *61*, 681 (1974)
7.33   M.J.L. Sangster, R.M. Atwood: J. Phys. C *11*, 1541 (1978)
7.34   J.M. Rowe, J.J. Rush, E. Prince, N.J. Chesser: Ferroelectrics *16*, 107 (1977)
7.35   P.R. Vijayaraghavan, S.K. Sinha, P.K. Iyengar in: *Annual Report of the Nuclear Physics Division,* Bhabha Atomic Research Center, Bombay (1976) p.69
7.36   P.R. Vijayaraghavan: private communication (1978)
7.37   R.J. Birgeneau, S.M. Shapiro in: *Valence Instabilities and Narrow-Band Phenomena,* ed. by R.D. Parks (Plenum Press, New York 1977) p.49
7.38   H.A. Mook, R.M. Nicklow, T. Penney, F. Holtzberg, W.M. Shafer: Phys. Rev. B *18*, 2925 (1978)
8.1   A.A. Ahmad, H.G. Smith, N. Wakabayashi, M.K. Wilkinson: Phys. Rev. B *6*, 3956 (1972)
8.2   J.F. Vetelino, K.V. Namjoshi, S.S. Mitra: Phys. Rev. B *7*, 4001 (1973)
8.3   A. Ghosh, A.N. Basu: J. Phys. C *9*, 4365 (1976)
8.4   M.J.L. Sangster, R.M. Atwood: J. Phys. C *11*, 1541 (1978)
8.5   S. Rolandson, G. Raunio: Phys. Rev. B *4*, 4617 (1971)
8.6   G. Mahler, P. Engelhardt: Phys. Status Solidi (b) *45*, 543 (1971)
8.7   J. Daubert in: *Neutron Inelastic Scattering* (IAEA Vienna 1972) p.85
8.8   H.H. Lal, M.P. Verma: J. Phys. C *5*, 1038 (1972)
8.9   J. Daubert, H. Jex, M. Müllner: Phys. Status Solidi (b) *57*, 477 (1973)
8.10   W. Bührer, W. Hälg: Phys. Status Solidi (b) *46*, 679 (1971)
8.11   J.S. Eldridge: J. Phys. C *3*, 1527 (1970)
8.12   B.I. Gorbachev, P.G. Ivanitskii, V.T. Krotenko, M.V. Pasechnik: Ukr. Fiz. Zh. (USSR) *18*, 92 (1973)
8.13   H.G. Smith in: G. Venkataraman, V.C. Sahni: Rev. Mod. Phys. *42*, 409 (1970)
8.14   H.Jex: Solid State Commun. *9*, 2057 (1971)
8.15   H.G. Smith, J.G. Traylor, W. Reichardt: Phys. Rev. *181*, 1218 (1969)
8.16   E.R. Cowley: Phys. Rev. B *3*, 2743 (1971)
8.17   C.H. Kim, H.A. Rafizadeh, S. Yip: J. Chem. Phys. *57*, 2291 (1972)
8.18   H.C. Teh: Can. J. Phys. *50*, 2807 (1972)
8.19   H.C. Teh.,B.N. Brockhouse: Phys. Rev. B *3*, 2733 (1971)
8.20   H. Jex: Phys. Lett. A *34*, 118 (1971)
8.21   H.C. Teh, B.N. Brockhouse: Phys. Rev. B *8*, 3928 (1973)
8.22   R.K. Singh, H.N. Gupta, M.K. Agrawal: Phys. Rev. B *17*, 894 (1978)
8.23   Y. Yamada, Y. Noda, J.D. Axe, G. Shirane: Phys. Rev. B *9*, 4429 (1974)
8.24   Y Fujii, T. Sakuma, J. Nakahara, S. Hoshino, K. Kobayashi, A. Fujii: J. Phys. Soc. Jpn. *44*, 1237 (1978)
8.25   E.R. Cowley, A. Okazaki: Proc. Roy. Soc. (London) A *300*, 45 (1967)
8.26   R. Srinivasan, G. Lakshmi, V. Ramachandran: J. Phys. C *8*, 2889 (1975)
9.1   J.L. Warren, J.L. Yarnell, G. Dolling, R.A. Cowley: Phys. Rev. *158*, 805 (1967)
9.2   R. Wehner, H. Borik, W. Kress, A. Goodwin, S.D. Smith: Solid State Commun. *5*, 307 (1967)
9.3   G. Dolling, R.A. Cowley: Proc. Phys. Soc. (London) *88*, 463 (1966)
9.4   G. Peckham: Solid State Commun. *5*, 311 (1967)
9.5   H.L. Mc Murry, A.W. Solbrig, J.K. Boyter, C. Noble: J. Phys. Chem. Solids *28*, 2359 (1967)
9.6   R. Blanchard, Y.P. Varshni: Phys. Rev. *159*, 599 (1967)
9.7   W. Cochran: Crit. Rev. Solid State Sci. *2*, 1 (1971)
9.8   S.K. Sinha: Crit. Rev. Solid State Sci. *3*, 273 (1973)

9.9   B.P. Pandey, B. Dayal: J. Phys. C *6*, 2943 (1973)
9.10   R. Tubino, L. Piseri, G. Zebri: J. Chem. Phys. *56*, 1022 (1972)
9.11   F.A. Johnson, K. Moore: Proc. Soc. (London) A *339*, 85 (1974)
9.12   W. Weber: Phys. Rev. B *15*, 4789 (1977)
9.13   G. Dolling: *Inelastic Neutron Scattering*, Vol. 2 (IAEA Vienna 1965)
       p.37
9.14   G. Nilsson, G. Nelin: Phys. Rev. B *6*, 3777 (1972)
9.15   B.N. Brockhouse: Phys. Rev. Lett. *2*, 256 (1959)
9.16   R.M. Martin: Phys. Rev. *186*, 871 (1969)
9.17   W. Solbrig, Jr.: Phys. Chem. Solids *32*, 1761 (1971)
9.18   D.L. Price, S.K. Sinha, R.P. Gupta: Phys. Rev. B *9*, 2573 (1974)
9.19   T. Soma: Phys. Status Solidi (b) *76*, 753 (1976)
9.20   G. Nilsson, G. Nelin: Phys. Rev. B *3*, 364 (1971)
9.21   G. Nelin, G. Nilsson: Phys. Rev. B *5*, 3151 (1972)
9.22   N.K. Pope in: *Lattice Dynamics*, ed. by R.F. Wallis (Pergamon Press,
       Oxford 1965) p.147
9.23   B.N. Brockhouse, P.K. Ivengar: Phys. Rev. *111*, 747 (1958)
9.24   F. Herman: J. Phys. Chem. Solids *8*, 405 (1959)
9.25   W. Cochran: Proc. Roy. Soc. (London) A *253*, 260 (1959)
9.26   W. Kress: Phys. Status Solidi (b) *49*, 235 (1972)
9.27   B.P. Pandy, B. Dayal: Solid State Commun. *11*, 775 (1972)
9.28   G. Nelin: Phys. Rev. B *10*, 4331 (1974)
9.29   W. Weber: Phys. Rev. Lett. *33*, 371 (1974)
9.30   A.M. Al'tschuler, Yu.Kh. Vekilo, T.I. Kacherets: Sov. Phys. Solid
       State *18*, 1279 (1976)
9.31   S.S. Jaswal in: *Lattice Dynamics*, ed. by M. Balkanski (Flammarion,
       Paris 1978) p.41
9.32   D.L. Price, J.M. Rowe, R.N. Nicklow: Phys. Rev. B *3*, 1268 (1971)
9.33   D.L. Price, J.M. Rowe: Solid State Commun. *7*, 1433 (1969)
9.34   R.P. Gupta, S.K. Sinha, J.P. Walter, M.L. Cohen: Solid State Commun.
       14, 1313 (1974)
9.35   T. Soma: J. Phys. Soc. Jpn. *36*, 1301 (1974)
10.1   K. Kunc, M. Balkanski, M.A. Nusimovici: Phys. Status Solidi (b) *72*,
       229 (1975)
10.2   D.W. Feldman, J.H. Parker, W.J. Choyke, L. Patrick: Phys. Rev. *173*,
       787 (1968)
10.3   J.F. Vetelino, S.S. Mitra: Phys. Rev. *178*, 1349 (1969)
10.4   R. Banerjee, Y.P. Varshni: J. Phys. Soc. Jpn. *30*, 1015 (1971)
10.5   T.N. Singh, S.S. Kushwaha, G. Singh: Solid State Commun. *13*, 1393
       (1973)
10.6   T.N. Singh, B.N. Roy: Nuovo Cimento B *41*, 198 (1977)
10.7   R. Banerjee, Y.P. Varshni: Can. J. Phys. *47*, 451 (1969)
10.8   L.L. Yarnell, J.L. Warren, R.G. Wenzel, P.J. Dean: *Neutron Inelastic
       Scattering*, Vol. 1, (IAEA, Vienna 1968) p.301
10.9   R. Banerjee, Y.P. Varshni: J. Phys. Soc. Jpn. *30*, 1015 (1971)
10.10 K. Kunc, M. Balkanski, M.A. Nusimovici: Phys. Status Solidi (b) *72*,
       229 (1975)
10.11 K. Kunc, H. Bilz in: *Proc. Conf. Neutron Scattering*, Vol. 1, ed. by
       R.M. Moon (Gatlinburg, USA 1976), CONF-760601-P1
       and Solid State Commun. *19*, 1027 (1976)
10.12 K.C. Rustagi, W. Weber: Solid State Commun. *18*, 1027 (1976)
10.13 J.L.T. Waugh, G. Dolling: Phys. Rev. *132*, 2410 (1963)
10.14 G. Dolling, R.A. Cowley: Proc. Phys. Soc. (London) *88*, 463 (1966)
10.15 E.N. Korol: Sov. Phys. Solid State *12*, 497 (1970)
10.16 M.K. Farr, J.G. Traylor, S.K. Sinha: Phys. Rev. B *11*, 1587 (1975)
10.17 M. Vandevyver, P. Plumelle: Phys. Rev. B *17*, 675 (1978)
10.18 P.H. Borcherds, G.F. Alfrey, D.H. Saunderson, A.D.B. Woods: J. Phys.
       C *8*, 2022 (1975)

10.19 M. Vandevyver, P. Plumelle: J. Phys. Chem. Solids *38*, 765 (1977)
10.20 P.H. Borcherds, K. Kunc: J. Phys. C *11*, 4145 (1978)
10.21 D.L. Price, J.M. Rowe, R.M. Nicklow: Phys. Rev. B *3*, 1268 (1971)
10.22 R. Banerjee, Y.P. Varshni: Can. J. Phys. *47*, 451 (1969)
10.23 D.N. Talwar, B.K. Agrawal: Solid State Commun. *11*, 1691 (1972)
10.24 E.S. Koteles, W.R. Datars, G. Dolling: Phys. Rev. B *9*, 572 (1974)
10.25 K. Kunc, M. Balkanski, M.A. Nusimovici: Phys. Rev. B *12*, 4346 (1975)
10.26 S.S. Jaswal in: *Lattice Dynamics*, ed. by M. Balkanski (Flammarion, Paris 1978) p.41
10.27 J. Bergsma: Phys. Rev. Lett. A *32*, 324 (1970)
10.28 L.A. Feldkamp, D.K. Steinman, N. Vagelatos, J.S. King, G. Venkataraman: J. Phys. Chem. Solids *32*, 1573 (1971)
10.29 J.F. Vetelino, S.S. Mitra, O. Brafman, T.C. Damen: Solid State Commun. *7*, 1809 (1969)
10.30 D.N. Talwar, B.K. Agrawal: Phys. Status Solidi (b) *63*, 441 (1974)
10.31 D.N. Talwar, B.K. Agrawal: Phys. Status Solidi (b) *64*, 71 (1974)
10.32 N. Vagelatos, D. Wehe, J.S. King: J. Chem. Phys. *60*, 3613 (1974)
10.33 B. Hennion, F. Moussa, G. Pepy, K. Kunc: Phys. Lett. A *36*, 377 (1971)
10.34 P. Plumelle, M. Vandevyver: Phys. Status Solidi (b) *73*, 271 (1976)
10.35 J.M. Rowe, R.M. Nicklow, D.L. Price, K. Zanio: Phys. Rev. B *10*, 671 (1974)
10.36 V.T. Bublik, S.S. Gorelik, I.S. Smirnow: Sov. Phys. Crystallogr. *17*, 485 (1972)
10.37 B. Prevot, B. Hennion, B. Dorner: J. Phys. C *10*, 3999 (1977)
10.38 C. Carabatos, B. Hennion, K. Kunc, F. Moussa, C. Schwab: Phys. Rev. Lett. *26*, 770 (1971)
10.39 Z. Vardeny, G. Gilat in: *Lattice Dynamics*, ed. by M. Balkanski (Flammarion, Paris 1978) p.55
10.40 S. Hoshino, Y. Fujii, J. Harada, J.D. Axe: J. Phys. Soc. Jpn. *41*, 965 (1976)
10.41 B. Prevot, C. Carabatos, C. Schwab, B. Hennion, F. Moussa: Solid State Commun. *13*, 1725 (1973)
10.42 B. Hennion, F. Moussa, B. Prevot, C. Carabatos, C. Schwab: Phys. Rev. Lett. *28*, 964 (1972)
11.1   G.L. Ostheller, R.E. Schmunk, R.M. Brugger, R.J. Kearnly: *Inelastic Neutron Scattering*, Vol. 1 (IAEA Vienna 1968) p.315
11.2   R.M. Brugger, K.A. Strong, J.M. Carpenter: J. Phys. Chem. Solids *28*, 249 (1967)
11.3   R. Ramani, K.K. Mani, R.P. Singh: Phys. Rev. B 14, 2659 (1976)
11.4   R.N. Sinclair: IAEA Symp. (Chalk River 1962), Vol. 2, p.199
11.5   A.W. Hewat: Solid State Commun. *8*, 187 (1970)
11.6   K. Thoma, B. Dorner, G. Duesing, W. Wegener: Solid State Commun. *15*, 1111 (1974)
11.7   W. Wegener, S. Hautecler: Phys. Lett. A *31*, 2 (1970)
11.8   K. Kunc, H. Bilz in *Proc. Conf. Neutron Scattering*, Vol. 1, ed. by R.M. Moon (Gatlinburg, USA 1976) CONF-760601-P1, p.195 and Solid State Commun. *19*, 1027 (1976)
11.9   M.A. Nusimovici, M. Balkanski, J.L. Birman: Phys. Rev. B *1*, 595 (1970)
11.10 W. Bührer, R.M. Nicklow, P. Bruesch: Phys. Rev. B *17*, 3362 (1978)
12.1   M.M. Elcombe, A.W. Pryor: J. Phys. C *3*, 492 (1970)
12.2   M. Elcombe: J. Phys. C *5*, 2702 (1972)
12.3   A. Sadoc, F. Moussa, G. Pepy: J. Phys. Chem. Solids *37*, 197 (1976)
12.4   J.P. Hurrell, V.J. Minkiewicz: Solid State Commun. *8*, 463 (1970)
12.5   M.H. Dickens, M.T. Hutchings: J. Phys. C *11*, 461 (1978)
12.6   R.J. Kearney, T.G. Worlton, R.E. Schmunk: J. Phys. Chem. Solids *31*, 1085 (1970)

12.7 N. Wakabayashi, A.A.Z. Ahmad, H.R. Shanks, G.C. Danielson: Phys. Rev. B 5, 2103 (1972)
12.8 G. Dolling, R.A. Cowley, A.D.B. Woods: Can. J. Phys. 43, 1397 (1965)
12.9 R.A. Cowley, G. Dolling: Phys. Rev. 167, 464 (1968)
13.1 J.G. Traylor, H.G. Smith, R.M. Nicklow, M.K. Wilkinson: Phys. Rev. B 3, 3457 (1971)
13.2 R.S. Katiyar in: Phonons, Proc. Int. Conf. Phonons, Rennes, ed. by M.A. Nusimovici (Flammarion, Paris 1971) p.84
13.3 R. Almairac, C. Benoit: J. Phys. C 7, 2614 (1974)
13.4 G.C. Cran, M.J.L. Sangster: J. Phys. C 7, 1937 (1974)
13.5 R.S. Katiyar: J. Phys. C 3, 1693 (1970)
13.6 B.D. Rainford, J.G. Houmann, H.J. Guggenheim: Neutron Inelastic Scattering (IAEA Vienna 1972) p.655
13.7 P. Martel, R.A. Cowley, R.W.H. Stevenson: Can. J. Phys. 46, 1355 (1968)
13.8 R.S. Katiyar, P. Dawson, M.M. Hargreave, G.R. Wilkinson: J. Phys. C 4, 2421 (1971)
13.9 R. Almairac, J.L. Sauvajol, C. Benoit, A.M. Bon: J. Phys. C 11, 3157 (1978)
14.1 M.R. Chowdhury, G.E. Peckham, R.T. Ross, D.H. Saunderson: J. Phys. C 7, L99 (1974)
14.2 C.H. Perry, H. Buhay, A.M. Quittet, R. Currat in: Lattice Dynamics, ed. by M. Balkanski (Flammarion, Paris 1978) p.677
14.3 R. Currat, R. Comes, B. Dorner, E. Wiesendanger: J. Phys. C 7. 2521 (1974)
14.4 R. Comes, G. Shirane: Phys. Rev. B 5, 1886 (1972)
14.5 J.D. Axe, J. Harada, G. Shirane: Phys. Rev. B 1, 1227 (1970)
14.6 G. Shirane: Rev. Mod. Phys. 46, 437 (1974)
14.7 R. Migoni, H. Bilz, D. Bäuerle: Phys. Rev. Lett. 37, 1155 (1976)
14.8 R. Migoni, H. Bilz in: Lattice Dynamics, ed. by M. Balkanski (Flammarion, Paris 1978) p.650
14.9 K. Gesi, J.D. Axe, G. Shirane, A. Linz: Phys. Rev. B 5, 1933 (1972)
14.10 V.J. Minkiewicz, G. Shirane: J. Phys. Soc. Jpn. 26, 674 (1969)
14.11 W.G. Stirling: J. Phys. C 5, 2711 (1972)
14.12 R.A. Cowley: Phys. Rev. 134, A 981 (1964)
14.13 Y. Yamada, G. Shirane: J. Phys. Soc. Jpn. 26, 396 (1969)
14.14 G. Shirane, Y. Yamada: Phys. Rev. 177, 858 (1969)
14.15 R.A. Cowley, W.J.L. Buyers, G. Dolling: Solid State Commun. 7, 181 (1969)
14.16 W.G. Stirling, R. Currat: J. Phys. C 9, L519 (1976)
14.17 R. Migoni, K.H. Rieder, K. Fischer, H. Bilz: Ferroelectrics 13, 377 (1976)
14.18 R.A. Cowley in: Lattice Dynamics, ed. by M. Balkanski (Flammarion, Paris 1978) p.625
14.19 A.J. Maeland: J. Chem. Phys. 51, 2915 (1969)
14.20 G. Shirane, B.C. Frazer, V.J. Minkiewicz, J.A. Leake: Phys. Rev. Lett. 5, 234 (1967)
14.21 Y.Yamada, G. Shirane, A. Linz: Phys. Rev. 177, 848 (1969)
14.22 J. Harada, J.D. Axe, G. Shirane: Acta Crystallogr. A 26, 608 (1970)
14.23 J. Harada, J.D. Axe, G. Shirane: Phys. Rev. B 4, 155 (1971)
14.24 G. Shirane, J.D. Axe, J. Harada: Phys. Rev. B 2, 3651 (1970)
14.25 J.D. Axe, G. Shirane, K.A. Müller: Phys. Rev. 183, 820 (1969)
14.26 G. Shirane, J.D. Axe, J. Harada, J.P. Remeika: Phys. Rev. B 2, 155 (1970)
14.27 B. Dorner, M. Steiner: J. Phys. C 9, 15 (1976)
14.28 T.M. Holden, W.J.L. Buyers, E.C. Svensson, R.A. Cowley, M.T. Hutchings, D. Hunkin, R.W.H. Stevenson: J. Phys. C 4, 2127 (1971)

14.29   E.R. Cowley, A.K. Pant: Phys. Rev. B *8*, 4795 (1973)
14.30   K.W. Logan, S.F. Trevino, R.C. Casella, W.M. Shaw, L.D. Mühlenstein,
        R.D. Mical in: *Phonons* (Proc. Int. Conf., Rennes), ed. by
        M.A. Nusimovici (Flammarion, Paris 1971) p.104
14.31   K.R. Rao, S.L. Chaplot, P.K. Iyengar, A.H. Venkatesch,
        P.R. Vijayaraghavan in: *Lattice Dynamics*, ed. by M. Balkanski
        (Flammarion, Paris 1978) p.73
14.32   K.R. Rao, S.L. Chaplot, P.K. Iyengar, A.H. Venkatesh,
        P.R. Vijayaraghavan: Pramana (India) *11*, 251 (1978)
15.1   R.M. Nicklow, N. Wakabayashi, H.G. Smith: Phys. Rev. B *5*, 4951
        (1972)
15.2   A. Yoshimori, Y. Kitano: J. Phys. Soc. Jpn. *11*, 352 (1956)
15.3   G. Dolling, B.N. Brockhouse: Phys. Rev. *128*, 1120 (1962)
15.4   J.A. Young, J.V. Koppel: J. Chem Phys. *42*, 357 (1965)
15.5   D.I. Page in: *Neutron Inelastic Scattering*, Vol. 1 (IAEA Vienna
        1968) p.325
15.6   A.P. Roy: Can. J. Phys. *49*, 277 (1971)
15.7   D.K. Ross: J. Phys. C *6*, 3525 (1973)
15.8   K.K. Mani, R. Ramani: Phys. Status Solidi (b) *61*, 659 (1974)
15.9   A.P.P. Nicholson, D.J. Bacon: J. Phys. C *10*, 2295 (1977)
15.10   B.N. Brockhouse, G. Shirane in: *Lattice Dynamics*, ed. by M. Balkanski
        (Flammarion, Paris 1978) p.123
15.11   N. Wakabayashi, H.G. Smith, R. Shanks: Phys. Lett. A *50*, 367
        (1974)
15.12   N. Wakabayashi: Nuovo Cimento B *38*, 256 (1977)
15.13   N. Wakabayashi, H.G. Smith, R.M. Niclow: Phys. Rev. B *12*, 659
        (1975)
15.14   J.L. Brebner, S. Jandl, B.M. Powell: Nuovo Cimento B *38*, 263
        (1977)
15.15   B. Dorner, R.E. Gosh, G. Harbecke: Phys. Status Solidi (b) *73*,
        655 (1976)
15.16   W.G. Stirling, B. Dorner, J.D.N. Cheeke, J. Revelli: Solid State
        Commun. *18*, 931 (1976)
15.17   D.E. Moncton, F.J. DiSalvo, J.D. Axe in: *Lattice Dynamics*, ed. by
        M. Balkanski (Flammarion, Paris 1978) p.561
15.18   A. Pasternak: J. Phys. C *9*, 2987 (1976)
15.19   W.B. Yelon, S. Scherm, C. Vettier: Solid State Commun. *15*, 391
        (1974)
15.20   A. Anderson, J.P. Todoeschuck: Can. J. Spectrosc. (Canada) *22*, 113
        (1977)
15.21   P. Carrara, J.P. Redoules, C. Escribe, K.R.A. Ziebeck: Solid State
        Commun. *21*, 929 (1977)
15.22   K. Wagner, G. Dolling, B.M. Powell, G. Landwehr: Phys. Status Solidi
        (b) *85*, 211 (1978)
15.23   H. Meyer, A. Weiss, B. Dorner: Solid State Commun. *25*, 1093 (1978)
15.24   B.M. Powell, S. Jandl. J.L. Brebner, F. Lêvy: J. Phys. C *10*, 3039
        (1977)
15.25   A. Polian, K. Kunc, R. Le Toullec, B. Dorner in: *Physics of
        Semiconductors, 1978*, ed. by B.L.H. Wilson (The Institute of
        Physics, London 1979) p.907
15.26   S. Jandl. J.L. Brebner, B.M. Powell: Phys. Rev. B *13*, 686 (1976)
15.27   W. Kress, A. Frey, B. Dorner, W. Kaiser in: *Proc. Conf. Neutron
        Scattering*, Vol. 1, ed. by R.M. Moon (Gatlinburg, USA 1976) p.216
15.28   H.J.L. van der Valk: Solid State Commun. *20*, 815 (1976)
15.29   A. Frey, R. Zeyher: Solid State Commun. *28*, 435 (1978)
15.30   D.E. Moncton, J.D. Axe, F.J. DiSalvo: Phys. Rev. B *16*, 801 (1977)
15.31   H.G. Smith, N. Wakabayashi in: *Dynamics of Solids and Liquids by
        Neutron Scattering*, ed. by T. Springer, S. Lovesey (Springer Berlin,
        Heidelberg, New York 1977)

16.1   W.D. Teuchert, R. Geick, G. Landwehr, H. Wendel, W. Weber: J. Phys.
       C 8, 3725 (1975)
16.2   J. Etchepare, P. Kaplan, M. Merian in: *Lattice Dynamics*, ed. by
       M. Balkanski (Flammarion, Paris 1978) p.60
16.3   B.A. Kotov, N.M. Okuneva, A.R. Regel, A.L. Shakh-Budagov: Sov. Phys.
       Solid State 9, 955 (1967)
16.4   R. Geick, U. Schröder, J. Stuke: Phys. Status Solidi 24, 99
       (1967)
16.5   T. Nakayama, A. Odajima: J. Phys. Soc. Jpn. 34, 732 (1973)
16.6   W. Hamilton, B. Lassier, M. Kay: J. Phys. Chem. Solids 35, 1089
       (1974)
16.7   H. Wendel, W. Weber, W.D. Teuchert: J. Phys. C 8, 3737 (1975)
16.8   B.M. Powell, P. Martel: J. Phys. Chem. Sclids 36, 1287 (1975)
16.9   W. Glisser, A. Axmann, T. Springer in: *Neutron Inelastic Scattering*,
       Vol. 1 (IAEA, Vienna 1968) p.245
16.10 S.A. Pine, G. Dresselhaus: Phys. Rev. B 4, 356 (1971)
16.11 E.R. Cowley: Can. J. Phys. 51, 843 (1973)
16.12 T.B. Gibbons: Phys. Rev. B 7, 1410 (1973)
16.13 H. Wendel: J. Phys. C 9, 445 (1976)
16.14 L. Pintschovius, R. Currat, H. Wendel in: *Lattice Dynamics*. ed. by
       M. Balkanski (Flammarion, Paris 1978) p.579
16.15 H. Wendel: J. Phys. C 10, L1 (1977)
16.16 T. Luthy, G.S. Pawley: Phys. Status Solidi (b) 69, 551 (1975)
16.17 W. Reichardt, K.H. Rieder in: *Proc. Conf. Neutron Scattering*, Vol. 1,
       ed. by R.M. Moon (Gatlinburg, USA 1976) p.181
16.18 G.A. Briggs, C. Duffill, M.T. Hutchings, R.D. Lowde,
       N.S. Satya-Murthy in: *Neutron Inelastic Scattering* (IAEA Vienna,
       1972) p.669
16.19 M.M. Beg, S.M. Shapiro: Phys. Rev. B 13, 1728 (1976)
16.20 C. Carabatos, B. Prevot: Phys. Status Solidi (b) 44, 701 (1971)
16.21 H. Prask, H. Boutin, S. Yip: J. Chem. Phys. 48, 3367 (1968)
16.22 T.H.K. Barron, C.C. Huang, A. Pasternak: J. Phys. C 9, 3925 (1976)
16.23 H. Grimm, H. Rzany, B. Dorner, H. Jagodzinski: Priv. commun. 1978
16.24 M.M. Elcombe: as cited in Ref. 16.22
16.25 M.M. Elcombe: Proc. Phys. Soc. (London) 91, 947 (1967)
16.26 M.E. Striefler, G.R. Barsch: Phys. Rev. B 12, 4553 (1975)
16.27 P. Tompson, N.W. Grimes: Solid State Commun. 25, 609 (1978)
16.28 M. Iszumi, J.D. Axe, G. Shirane, K. Shimaoka: Phys. Rev. B 15, 4392
       (1977)
16.29 R.P. Rinaldi, G.S. Pawley: J. Phys. C 8, 599 (1975)
16.30 B.A. Kortov, N.M. Okuneva, A.L. Shakh-Budagov: Sov. Phys. Solid
       State 9, 2011 (1968)
16.31 J. Skalyo, B.C. Frazer, G. Shirane: Phys. Rev. B 1, 278 (1970)
16.32 P. Bosi, R. Tubino, G. Zerbi: J. Chem. Phys. 59, 4578 (1973)
16.33 B. Renker: Phys. Lett. A 30, 493 (1969)
17.1   J.K. Kjems, G. Dolling: Phys. Rev. B 11, 1639 (1975)
17.2   A. Pasternak, A. Anderson, J.W. Leech: J. Phys. C 10, 3261 (1977)
17.3   H.G. Smith, M. Nielsen, C.B. Clark: Chem. Phys. Lett. (Netherlands)
       33, 75 (1975)
17.4   B.M. Powell, G. Dolling, L. Piseri, P. Martel in: *Neutron Inelastic
       Scattering* (IAEA, Vienna 1972) p.207
17.5   P.S. Goyal, B.A. Dasannacharya, C.L. Thaper, P.K. Iyengar: Phys.
       Status Solidi (b) 50, 701 (1972)
17.6   S.F. Trevino, M.K. Farr, P.A. Giguera, J.L. Arman in: *Proc. Conf.
       Neutron Scattering*, ed. by R.M. Moon (Gatlinburg, USA 1976) p.152
17.7   P. Bruesch, M. Bösch, W. Känzig, M. Ziegler, W. Bührer: Phys. Status
       Solidi (b) 77, 153 (1976)

17.8   K.R. Rao, S.F. Trevino, H. Prask: Phys. Rev. B *4*, 4551 (1971)
17.9   I.U. Heilmann, N.B. Olsen, D.J. Lockwood, G.A. Mackenzie,
        G.S. Pawley in: *Lattice Dynamics*, ed. by M. Balkanski (Flammarion,
        Paris 1978) p.475
17.10   J. Sakurai, R.A. Cowley, G. Dolling: J. Phys. Soc. Jpn. *28*, 1426
        (1970)
18.1   L.K. Vodopyanov, I.V. Kutcherenko, A.P. Shotov, R. Scherm in:
        Lattice Dynamics, ed. by M. Balkanski (Flammarion, Paris 1978)
        p.673
18.2   W.J. Daughten, E. Gürmen, C.W. Tompson in: *Proc. Conf. Neutron
        Scattering*, Vol. 1, ed. by R.M. Moon, (Gatlinburg, USA 1976) p.145
18.3   C.H. Perry, I.R. Jahn, V. Wagner, W. Bauhofer, L. Genzel,
        J.B. Sokoloff in: *Lattice Dynamics*, ed. by M. Balkanski (Flammarion,
        Paris 1978) p.419
18.4   Y. Yamada, Y. Noda, J.D. Axe, G. Shirane: Phys. Rev. B *9*, 4429
        (1974)
18.5   L. Genzel, T.P. Martin, C.H. Perry: Phys. Status Solidi (b) *62*, 83
        (1974)
18.6   H.G. Smith, J.G. Traylor, W. Reichardt: Phys. Rev. *181*, 1218
        (1969)
18.7   W.J.L. Buyers, R.A. Cowley in: *Inelastic Neutron Scattering*, Vol. 1
        (IAEA, Vienna 1968) p.43
18.8   J. Asam, S. Rolandson, M.M. Beg. N.M. Butt, Q.H. Khan: Phys. Status
        Solidi (b) *77*, 693 (1977)
18.9   N.E. Massa, J.F. Vetelino, S.S. Mitra in: *Lattice Dynamics*, ed. by
        M. Balkanski (Flammarion, Paris 1978) p.425
19.1   W.G. Stirling, W. Press, H.H. Stiller: J. Phys. C *10*, 3959 (1977)
19.2   B. Dorner, H. Stiller: Phys. Status Solidi *18*, 795 (1966)
19.3   H. Stiller, S. Hautecler in: *Inelastic Scattering of Neutrons in
        Solids and Liquids*, Vol. 2, (IAEA, Vienna 1963) p.281
19.4   B. Dorner, H. Stiller: Phys. Status Solidi *5*, 511 (1964)
19.5   W. Press, B. Dorner, H. Stiller: Solid State Commun. *9*, 1113 (1971)
19.6   P.A. Reynolds: J. Chem. Phys. *59*, 2777 (1973)
19.7   P.A. Reynolds, J.K. Kjems, J.W. White: J. Chem. Phys. *60*, 824 (1974)
19.8   G. Dolling, B.M. Powell: Proc. Roy. Soc. A *319*, 209 (1970)
19.9   L.N. Becka: J. Chem. Phys. *37*, 431 (1962)
19.10   W. Bührer, W. Hälg, T. Schneider in: Rep. Inst. Reaktorforsch.,
        AF-SSP-9, Würenlingen 1967
19.11   G. Dolling, B.M. Powell, G.S. Pawley: Proc. Roy. Soc. A *333*, 368
        (1973)
19.12   T. Wasiutynski: Phys. Status Solidi (b) *76*, 175 (1976)
19.13   P. Martel, B.M. Powell, L.A. Vinhas in: *Lattice Dynamics*, ed. by
        M. Balkanski, (Flammarion, Paris 1978) p.494
19.14   P. Martel, B.M. Powell: Chem. Phys. Lett. *39*, 339 (1976)
19.15   E.L. Bokhenkov, E.F. Sheka, B. Dorner, I. Nathaniec: Solid State
        Commun. *23*, 89 (1977)
19.16   E.L. Bokhenkov, E.F. Sheka, I. Nathaniec, B. Dorner, W. Drexel in:
        Lattice Dynamics, ed. by M. Balkanski, (Flammarion, Paris 1978)
        p.471
19.17   A.I. Kitaigorodski: J. Chem. Phys. *63*, 9 (1966)
19.18   D.E. William: J. Chem. Phys. *45*, 3370 (1966); *47*, 4680 (1967)
19.19   G.S. Pawley, E.A. Yeats: Solid State Commun. *7*, 385 (1969)
19.20   E.L. Bokhenkov, I. Nathaniec, E.F. Sheka: Sov. Phys. JETP *43*, 536
        (1976)
19.21   G.A. Mackenzie, G.S. Pawley, O.W. Dietrich: J. Phys. C *10*, 3723 (1977)
19.22   U.A. Lutz, W. Halg: Solid State Commun. *8*, 165 (1970)
19.23   J. Lefebvre, M. More, R. Fouret, B. Hennion, R. Currat: J. Phys. C *8*,
        2011 (1975)

Subject Index

J. L. Birman
Licht und Materie I b
Light and Matter I b

Editor: L. Genzel
1974. 34 figures. XVI, 538 pages
(Handbuch der Physik, Gruppe 5,
25. Band, 2. Teil b)
ISBN 3-540-06638-1

"...The choice of what is important and the development of the theme from the general to the very particular suited me admirably, and I feel sure that solid-state physicists will benefit greatly from this part."   *Contemporary Physics*

"...an excellent article of encyclopedia..."
                                Physics Bulletin

Light Scattering in Solids

Editor: M. Cardona
1975. 111 figures, 3 tables. XIII, 339 pages
(Topics in Applied Physics, Volume 8)
ISBN 3-540-07354-X

Contents:
M. Cardona: Introduction. – *A. Pinczuk:*
E. Burstein: Fundamentals of Inelastic Light Scattering in Semiconductors and Insulators. – *R. M. Martin, L. M. Falicov:*
Resonant Raman Scattering. – *M. V. Klein:*
Electronic Raman Scattering. –
M. H. Brodsky: Raman Scattering in Amorphous Semiconductors. – *A. S. Pine:*
Brillouin Scattering in Semiconductors. –
Y.-R. Shen: Stimultated Raman Scattering

Springer-Verlag
Berlin
Heidelberg
New York

Photoemission in Solids I

General Principles

Editors: M. Cardona, L. Ley
1978. 90 figures, 17 tables. XI, 290 pages
(Topics in Applied Physics, Volume 26)
ISBN 3-540-08685-4

Contents:
M. Cardona, L. Ley: Introduction. –
W. L. Schaich: Theory of Photoemission:
Independent Particle Model. –
S. T. Manson: The Calculation of Photo-ionization Cross Sections: An Atomic View. – *D. A. Shirley:* Many-Electron and Final State Effects: Beyond the One-Electron Picture. – *G. K. Wertheim,*
P. H. Citrin: Fermi Surface Excitations in X-Ray Photoemission Line Shapes From Metals. – *N. V. Smith:* Angular Dependent Photoemission.

Photoemission in Solids II

Case Studies

Editors: L. Ley, M. Cardona
1979. 214 figures, 26 tables.
XVIII, 401 pages
(Topics in Applied Physics, Volume 27)
ISBN 3-540-09202-1

Contents:
L. Ley, M. Cardona: Introduction. – *L. Ley,*
M. Cardona, R. A. Pollak: Photoemission in Semiconductors. – *S. Hüfner:* Unfilled Inner Shells: Transition Metals and Compounds. – *M. Campagna, G. K. Wertheim,*
Y. Baer: Unfilled Inner Shells: Rare Earths and Their Compounds. – *W. D. Grobman,*
E. E. Koch: Photoemission from Organic Molecular Crystals. – *C. Kunz:*
Synchrotron Radiation: Overview. –
P. Steiner, H. Höchst, S. Hüfner: Simple Metals.

W. Ludwig

Recent Developments in Lattice Theory

1967. 87 figures. VI, 301 pages
(Springer Tracts in Modern Physics,
Volume 43)
ISBN 3-540-03982-1

Contents:
Introduction. – Symmetry and Invariance
Properties of the Coupling Parameters. –
Different Models for the Potential Energy
in Crystals. – Dynamics of Molecular
Crystals. – Anharmonic Effects in
Thermodynamical Properties. – Point
Defects in Crystal Lattices. – Interaction
of Phonons with Particles and Radiation. –
Appendix. – References.

Excitions

Editor: K. Cho
1979. 118 figures, 4 tables. Approx.
290 pages
(Topics in Current Physics, Volume 14)
ISBN 3-540-09567-5

Contents:
K. Cho: Introduction. – *K. Cho:* Internal
Structure of Excitions. – *P. J. Dean,
D. C. Herbert:* Bound Excitions in Semi-
conductors. – *B. Fischer, J. Lagois:* Surface
Exciton Polaritons. – *P. Y. Yu:* Study of
Excitions and Exciton-Phonon Inter-
actions by Resonant Raman and Brillouin
Spectroscopies.

G. Leibfried, N. Breuer

Point Defects in Metals I

Introduction to the Theory

1977. 138 figures, 22 tables. XIV, 342 pages
(Springer Tracts in Modern Physics,
Volume 81)
ISBN 3-540-08375-8

Contents:
Introduction and Survey. – Harmonic
Approximation and Linear Response
(Green's Function) of an Arbitrary
System. – Lattice Theory. – Continuum
Theory. – Transition From Lattice to
Continuum Theory. – Statistics and
Dynamics of Simple Single Point Defects. –
Scattering of Neutrons and X-Rays by
Crystals. – Probability, Distributions and
Statistics. – Properties of Crystals With
Defects in Small Concentration. –
Appendix.

Springer-Verlag
Berlin
Heidelberg
New York